AF697072

WITHDRAWN

Russian workers

MANCHESTER
UNIVERSITY PRESS

Russian workers

The anatomy of patience

SARAH ASHWIN

Manchester University Press
MANCHESTER AND NEW YORK

distributed exclusively in the USA by St. Martin's Press

Copyright © Sarah Ashwin 1999

The right of Sarah Ashwin to be identified as the author of this work has been asserted by her in accordance with the Copyright, Designs and Patents Act 1988.

Published by Manchester University Press
Oxford Road, Manchester M13 9NR, UK
and Room 400, 175 Fifth Avenue, New York, NY 10010, USA
http://www.man.ac.uk/mup

Distributed exclusively in the USA by
St. Martin's Press, Inc., 175 Fifth Avenue, New York,
NY 10010, USA

Distributed exclusively in Canada by
UBC Press, University of British Columbia, 6344 Memorial Road,
Vancouver, BC, Canada V6T 1Z2

British Library Cataloguing-in-Publication Data
A catalogue record for this book is available from the British Library

Library of Congress Cataloging-in-Publication Data applied for

ISBN 0 7190 5611 X *hardback*

First published 1999

06 05 04 03 02 01 00 99 10 9 8 7 6 5 4 3 2 1

HD
6735.15
.A84
1999

Typeset in Photina
by Servis Filmsetting Ltd, Manchester
Printed in Great Britain
by Bookcraft (Bath) Ltd, Midsomer Norton

Contents

Preface

The collapse of the rouble in August 1998 revealed just how misguided was talk of the forthcoming Russian boom. Not only had the neo-liberal reform strategy precipitated the longest and deepest recession in recorded world history, it had also noticeably failed to foster democratisation. Communism was replaced by the 'market Bolshevik' practice of the reformers and their sponsors in the international financial institutions, according to which the creation of a market economy took priority over democratic procedures, the rule of law and the immediate well-being of the population.[1] Russian workers in particular were encouraged to make sacrifices for a golden future which now seems further away than it ever did.

Although there have been sporadic protests in the wake of the financial collapse, Russians have generally responded to the crisis of 1998 with a despairing resignation, viewing it as merely the latest in a long line of 'deceptions' perpetrated by the ruling elite. The customary demonstrations on the eve of the October Revolution, for example, attracted fewer protestors in 1998 than in previous years. The trade-union-organised Day of Collective Action on 7 October was far larger, but even on the trade unions' own estimates the crowds were much smaller than expected. Far from revealing that Russians have nothing to protest about, this apathy merely highlights the depth of cynicism with which ordinary Russians regard the political process, and their sense that it is not within their power to influence the future direction of their country.

This pervasive feeling of impotence is self-perpetuating: because ordinary Russians feel that politics is beyond their control, they usually do not even attempt to influence the decisions affecting their lives. Nowhere is this more

[1] The most dramatic examples of this tendency were President Yeltsin's military assault on the Russian Parliament in 1993 because of its resistance to reform, and the institutionalisation of non-payment of wages and pensions as a routine form of economic management at all levels. Both of these practices were tacitly accepted by Western governments and the international financial institutions as necessary evils of the reform era.

apparent than in the sphere of industrial relations. Workers have done little to challenge the practices of enterprise, local and state officials and politicians which have, among other things, left them without wages for months at a time.

This apparent passivity was a problem that I constantly encountered while working as research officer for the Central and East European desk of the International Confederation of Free Trade Unions between 1991 and 1993. Working with Russian trade unions was a fascinating but frustrating process. This frustration is well captured by the comment of one of my former colleagues, a charismatic North American trade union organiser, used to working in the hostile environment of Texas, who had hoped to engender some worker activism in Russia. After an initial period of optimism he declared himself beaten: 'Oh man, it's impossible, they're like slaves.' The experience of this period provoked a number of questions. Why was the representation of workers' interests so problematic in post-Soviet Russia even after the substantial liberalisation that occurred in the late Gorbachev era? What prevented the former communist trade unions, which organised the overwhelming majority of Soviet workers in the past, from representing workers' interests? And, given the ineffectiveness of their official representatives, why did workers not organise outside the ranks of the official unions to defend their interests?

My analysis is based on an ethnographic case study of a mine located in Russia's largest and most important coal field, the Kuzbass in Western Siberia. The workers of the coal-mining industry, once celebrated by the Communist Party as heroes of the Soviet Union, have been at the forefront of the struggle over economic reform, with the World Bank pressing for a restructuring of the industry involving massive job losses, in particular in the Kuzbass where the Bank originally wanted employment in coal mining reduced by over 70 per cent. I decided to focus on mineworkers because they are one of the few groups of workers who have shown themselves capable of mounting significant national protests, but even so their response to restructuring has been piecemeal and ineffective. This reflects the fact that the basis of effective collective action – workers' self-organisation – is lacking even among the most active Russian workers. In this book I set out to explore the nature of the barriers to workers' organisation through an analysis of the lives of concrete individuals, their work groups and community, relating workers' political behaviour to the form and content of their everyday lives.

This focus on the micro-level is important precisely because it is here that democracy must take root if Russia is not to endure yet more authoritarian rule. A key lesson of the collapse of the neo-liberal project in Russia is that individual 'true believers' in government cannot be relied upon to deliver reform. The creation of an effective trade union movement in Russia (whether through the renewal of the former communist trade unions or through the growth of new organisations) would be a huge step towards making institutions and officials at all levels accountable to the people they are supposed to serve, and thus towards the establishment of democracy in Russia. This book is therefore dedicated to all those people working to secure a role for trade unionism in post-communist Russia.

Sarah Ashwin
November, 1998

Acknowledgements

My research would have been impossible to conduct without the assistance I received from the sociologists Olga Pulyaeva and Kostya Burnyshev in the Kuzbass. They arranged the initial access to the mines I visited and helped me find my way to them; gave indispensable advice about local politics and productive lines of questioning; had me to stay at their flats; and endured long evenings during which I talked compulsively about trade unions – a subject of which I am sure they were sick by the time I left! Petr Bizyukov, Veronika Bizyukova, Inna Donova and Lena Varshavskaya were also very helpful during my visits to the Kuzbass.

The people of 'Vishnovka', the settlement where I conducted my main case study, overwhelmed me with their generosity and open-heartedness. I would like to thank all those I interviewed during the research, but I am particularly grateful to those with whom I became close friends. Valya Kashtueva was the first worker to befriend me. I interviewed her while she was at work, and she immediately invited me to spend the weekend at her house. On meeting me her mother said, 'I'm so proud that you have chosen to visit my daughter . . . A foreigner in our house!' Luckily for me, having heard this story other generous Vishnovkans decided that they too would like a visit from a foreigner. As a result my life has been permanently enriched by my friendships with Valentina Ivanovna, Lyuda Zlobina, Igor Zlobin, Lyuba Sinkina, Raisa Zaitseva, and the irrepressible and indefatigable Yakov Ivanovich. I would also like to thank members of the trade union committee at Taldym for their time and openness.

Michael Burawoy's comments on the pieces of my work he has seen have been very stimulating and helpful. Meanwhile, my discussions with the sociologists Marina Kiblitskaya and Volodya Ilyin have always been fascinating and enjoyable. Both of them helped me to develop my ideas, and I am very grateful to them for their friendship and assistance.

This book developed out of my Ph.D. thesis, with which I received invaluable assistance from my supervisor, Simon Clarke. I would like to thank him for his support, generosity and inspiration, and for making this project so enjoyable. I am also indebted to my examiners, Huw Beynon and Robin Cohen, for their comments and assistance. I also wish to thank Lin Clarke for her friendship and support while I was completing my Ph.D.

The weaknesses which remain after all this assistance are entirely my responsibility.

Some of the chapters contain material from my previously published work. Chapter 1 contains material originally published in S. Ashwin (1998), 'Endless Patience: Explaining Soviet and Post-Soviet Social Stablility', *Communist and Post-Communist Studies*, 31, 2, 187–98. Chapter 2 contains some sections originally published in S. Ashwin (1995), '"There's No Joy Any More": The Experience of Reform in a Kuzbass Mining Settlement', *Europe–Asia Studies*, 47, 8, pp. 1267–381. Material from S. Ashwin (1997), 'Shop Floor Trade Unionism in Russia: The Prospects of Reform From Below', *Work, Employment and Society*, 11, 1, 115–31, has been used in Chapters 5 and 7. Chapter 6 contains some material which originally appeared in S. Ashwin (1996), 'Forms of Collectivity in a Non-Monetary Society', *Sociology*, 30, 1, 21–39, while Chapters 7 and 8 employ material from S. Ashwin (1998), 'Redefining the Collective: Russian Mineworkers in Transition from Communism', in M. Burawoy and K. Verdery (eds), *Uncertain Transition* (Boulder, Colorado, Rowman and Littlefield). Thanks to the editors and publishers for permission to use this work.

1

The anatomy of patience: Russian workers in transition from communism

> I am pessimistic about the future. I will not stand for trade union president again when my term comes to an end. . . . It's hopeless. In order to get a Russian to react you would literally have to slit his mother's throat, or kill his children right in front of him: only then might you stir him into action.

This was the lament one of the most militant mine trade union presidents in the Kuzbass in June 1998. His members, about whom he feels so despairing, have shown themselves to be some of the most active in Russia. Their mine, Taldym,[1] the subject of the case study on which this book is based, was one of the first mines to come out on strike in April 1989, before the famous protest of July that year. Moreover, the union president's despair was uttered in the immediate aftermath of the miners' action of May 1998, during which Russia was split in two by miners' blockades of the Trans-Siberian and North Caucasus railways mounted in protest at wage arrears. The miners appeared to emerge as the winners in this struggle: although their calls for President Boris Yeltsin's resignation and a reversal of economic policy had no impact, they were placated with part-payment of the wage debt and a schedule for full repayment. So why was the trade union president so despondent?

His despondency, and that of most activists, stems from the fact that the 'social explosion' they have been predicting for nearly ten years has failed to materialise despite the catastrophic decline in workers' living standards. Enterprises can delay payment of wages for six months or more, and workers will still turn up for work, only finally taking action when it becomes almost impossible for them to survive any longer without wages. Thus, in place of the much-predicted social explosion there has been a whole series of local protests. When wage delays have reached intolerable levels, for example, workers have, like the miners in May 1998, blocked railways or bridges or motorways, held hunger strikes and taken hostages.[2] There have even been cases when, in their

[1] Names have been changed to protect informants.

[2] Although the events of May 1998 received wide coverage in the Western press, the tactic is not new. The Anzhero-Sudzhensk miners blocked the Trans-Siberian in October 1994, and have

desperation to draw attention to their plight and that of their communities, workers have committed suicide in protest.[3] Such events do not count as *the* social explosion because, like the protest of May 1998, they do not destabilise the government or derail the economic reform programme. At best, they achieve temporary relief for those involved, though in many cases such action is merely ignored. This explains the depression of the trade union president: he was well aware that the government would not fulfil its promises and that wage debts would begin to mount again, as indeed occurred. The question this book addresses is therefore not so much why workers do not protest – in extreme circumstances they do – but why the action that they do take is so ineffective. The central issue in this regard is workers' organisation. Workers are not organised and such action as they do take does not leave an organisational legacy.

The trade union president was equally depressed about the inactivity of his union, Rosugleprof, and in this he is also not alone. The former communist branch trade unions, which, earlier on in the transition, some commentators thought might lead the disaffected in opposition to reform,[4] have been supine in the face of the change. They have not proved able to channel and represent workers' grievances during the post-communist period. This book attempts to explain both the failure of the official trade unions and the reasons why workers have proved unable either to reform the former communist trade unions from below or to form their own organisations.

The introduction begins with an analysis of the form of workers' integration within the Soviet polity and the role of the trade unions within the system. Anticipating the more detailed argument of the book as a whole, this chapter shows that an analysis of the forms of workers' integration within the enterprise makes it possible to explain the anatomy of 'patience' in the transition period. It also examines the structural position of trade unions under communism, laying the foundations for the analysis of their role during transition. Both the form of workers' integration and the position of trade unions are

since resorted to the tactic whenever wage delays become intolerable (for details see Borisov *et al.*, 1996). In other parts of the country, bridges or roads are the chosen site of protest. A typical example was the action of the workers of the 'Aviastar' factory in Ul'yanovsk, who blocked the bridge across the Volga for eight hours on 2 September 1996 in protest at a five-month wage delay, only ceasing their protest once the director promised to pay them (*Rabochaya sila*, 5: 23, September 1996: 2). In areas where there is no site for a sufficiently disruptive blockade, workers can become even more desperate in their attempts to gain the attention of the authorities. In the Kuzbass town of Prokop'evsk, for example, which spent the whole of winter 1996–7 without wages and heating, the teachers resorted to taking the chief of the education department hostage (Clarke *et al.*, 1997).

[3] The number of public suicides has been growing since 1997. For example, in the run-up to the May 1998 protest there was a wave of suicides in Rostov, where one miner hanged himself and another threw himself down the mine shaft in drastic pleas for wage debts to be paid to their families (ICFTU, 1998).

[4] The most prominent proponent of this point of view was Leonid Gordon, who argued, along with E. Klopov, that 'The old unions, in which egalitarian ideology is strong, will probably become the articulators of the interests of groups of workers least prepared for the market' (Gordon and Klopov, 1991: 41).

located within a comparative framework, in order to highlight their specific features. This is followed by a methodological discussion which explains the choice of Taldym as a case study, as well as the basis on which the study's findings can be generalised. The chapter concludes with an outline of the structure of the book.

Explaining soviet social stability

Understanding the mechanisms through which workers' compliance was secured in the past is crucial to explaining their behaviour in the face of reform. Various arguments have been developed to explain how the regime maintained social peace once it abandoned mass terror, in particular the theories of a 'social contract', 'incorporation' or 'atomisation'. The problem with these theories is that the various features of Soviet politics and society that they identify as being the keys to explaining social stability have all been destroyed along with the communist political system, and yet workers have remained quiescent. These theories' failure to account for the social peace of the transition era suggests that none of them has adequately conceptualised the form of workers' integration under communism. This section will examine the established accounts of Soviet social stability and highlight the problems with these ideas that have been exposed by workers' response to transition.

The most commonly employed explanation of workers' integration has been that a 'social contract' existed between the regime and the working class. According to the social contract thesis, workers offered their compliance in return for security of employment, stable prices, gradual improvement in consumption levels, and a relatively egalitarian system of remuneration, and because the Party allowed them 'considerable personal freedom as workers and consumers' (Hauslohner, 1987: 58).[5] The 'social contract' was expressed in terms of acceptance by both parties of certain rules of the game, departure from which was seen as unacceptable. These rules were institutionalised in laws and decrees and were guaranteed by the Labour Ministry and trade unions. Although the subscribers to the social contract thesis do allow that repression existed, they play down its significance, and assert that the stability of the system rested on 'volunteered (rather than coerced) compliance' (Hauslohner,

[5] Linda J. Cook (1993) lists the following authors as subscribing to the idea of a 'trade off of social welfare benefits for workers' quiescence' (219): Peter Hauslohner (1987); Walter Connor (1988: 67–85); Ed A. Hewett (1988: 39–50); Seweryn Bialer (1980: 158–65); Gail Lapidus (1983: 188–92) and George Breslauer (1984: 220–2). Cook claims that the social contract has three main empirical requirements: that the Soviet regime consistently deliver economic security and social welfare to workers; that it does so because it is constrained by its perception of workers' expectations; and that workers in return give their 'political compliance and quiescence' (Cook, 1993: 5). The question which is dealt with here is not whether the regime actually provided the benefits proponents of the social contract thesis claim, but whether the compliance of workers can be adequately explained by the idea that they were 'bought off' by those benefits which they did receive.

1987: 57). This view is very similar to that of David Lane, who argues that Soviet workers were 'incorporated' through permanence of employment; rising living standards; the organisation of enterprises into collectives providing social services to workers; toleration of slack labour discipline; the fact that managers and workers were socially closer than in the West and workers appreciated strong managers; and the incorporation of trade unions and 'natural leaders' among workers who were promoted or given a role in the Party or trade union (Lane, 1985: 165). Lane does also note that the 'threat of sanctions' was effective in binding the workers to the system, but he again places the emphasis on worker consent rather than regime manipulation, claiming that 'inputs of mass loyalty are not absent in the Soviet Union' (166).

In opposition to such views, authors such as Viktor Zaslavsky (1982) and Don Filtzer (1992, 1994)[6] have stressed that it was the regime-propagated 'atomisation' of the working class which explains the lack of resistance in the post-Stalin era. For Filtzer, the crucial determinant of workers' behaviour was their inability to organise collectively. They were 'atomised to an extreme degree' because they had no means of influencing political events through parties or trade unions; because political opposition of an individual or collective nature was prevented by a 'ubiquitous and brutal secret police', which quickly crushed incipient organisation; and because of the 'hyper-individualization of the labour process and incentives', which 'reinforced the disintegration of workers' solidarity' (Filtzer, 1992: 224–7). Therefore, the only avenue open to workers was to develop individualised forms of resistance – absenteeism, drunkenness, pilfering enterprise property, insubordination, job changing and control over production – which the system could tolerate because the goods produced were not 'commodities seeking realisation as exchange value'.[7] Meanwhile, Zaslavsky also sees job security and rising living standards as only part of the explanation for the lack of collective resistance on the part of workers. He again emphasises the political monopoly of the Communist Party and the threat of repression, while also highlighting the regime's deliberate segmentation of the working class through the creation of different status groups within it, and the safety valve offered by individual survival strategies such as job changing.

Both ways of seeing the relationship between workers and the regime contain important elements of the truth. First, the material security of Soviet workers did play an important role in ensuring political quiescence: the fact that strikes tended to occur in reaction to price rises or piece rate revisions bears eloquent testimony to the fact that workers did expect a certain minimal standard of living.[8] Meanwhile, proponents of the atomisation thesis are right to

[6] Filtzer's argument was influenced by the analysis of Soviet society developed by Hillel Ticktin. See Ticktin, 1992.

[7] That is, enterprises were not required to make a profit, but only to make the plan.

[8] W. D. Connor has used the evidence of strikes to argue for the existence of a social contract. He notes that there was a wave of strikes in 1962, the year of nationwide price increases. After

stress the role of repression and the regime's divisive tactics. The role of fear cannot be minimised: as late as 1989, when the miners embarked on the USSR's first industry-wide strike, one strike leader claimed that, 'Everyone had an instinctive fear that the strike might be suppressed by military force' (Clarke *et al.*, 1995: 71).[9]

Despite these merits, however, the experience of the reform era has exposed serious weaknesses in the accounts outlined above. Problems were already apparent in the social contract thesis before the collapse of communism, such as the uneven fulfilment of the putative 'contract' between different regions, sectors and categories of worker as well as between men and women.[10] It was hard to believe, for example, that low-paid female textile workers living in dormitories, and many other categories of workers, were 'bought off' by the beneficence of their communist masters.[11] But what has completely undermined the explanatory power of the theory is the events of the post-communist period. If the key to Soviet social stability was material security, why did workers not react a good deal more strongly to the flagrant violation of all aspects of the 'contract' during the economic reform period? The atomisation thesis, however, faces equally serious problems. If fear and repression played such a crucial role in maintaining social peace, why did a workers' movement not emerge once fear of reprisals had subsided after the miners' strike of 1989? This question is particularly pertinent given all the provocations proffered by the economic reformers.

A more recent approach to explaining the integration of Soviet workers,

these strikes the regime realised that price stability was an important element of the 'social contract' and did not openly raise the prices of basic foods again until the Gorbachev era. The fact that strikes occurred in reaction to any erosion of living standards has led Connor to conclude that 'Brezhnev era strikers saw themselves as standing up against violations of a just social contract' (Connor, 1991: 221).

[9] Rumours of past repression certainly aided the regime: in 1989 the miners were aware of the demonstration which had been put down by force in Novocherkassk in 1962, killing an estimated 70–80 demonstrators (Clarke *et al.*, 1995: 28).

[10] The inequalities between different categories of worker call into question the idea that workers were rewarded for their compliance. It seems more likely that the provision of rewards was directed towards addressing problems of labour motivation and discipline, which would explain why the priority sectors of the economy received the best treatment. The favourable treatment received by Moscow and Leningrad does suggest that Cook is right to claim that the regime was to some extent 'constrained by its . . . fear of labor discontent' (1993: 5), but increasing production was also a key concern.

[11] One answer to such criticism is to stress the fact that the Soviet labour force was 'unsophisticated' (Lane, 1985: 166). It is the case that a major point of comparison in the pre-Gorbachev era was the Soviet past rather than 'the West', but this does not necessarily imply that workers were satisfied with their lowly lot. Another point of comparison was Soviet managers who had access to closed stores, decent housing and monopolised enterprise leisure facilities. The common complaint of Russian workers when interviewed – that it is only such managers and those connected to them who are granted the privilege of living 'like human beings' (*po chelovecheskii*) – potently reveals that however 'unsophisticated' they may have been, workers were not satisfied by the minimal material provision on offer to them under communism.

developed with regard to the Stalinist era, has drawn on the tradition of post-structuralist class analysis popularised by social historians such as Gareth Stedman Jones. In this form of analysis, the articulation of 'languages of class' is accorded a key role in the achievement of class identification and action. Indeed, in its most radical formulation this 'discursive' approach almost reduced class to a linguistic creation: Stedman Jones came to reject what he termed 'essentialist' Marxist and sociological definitions of class based on a socio-economic logic (Stedman Jones, 1983: 21), treating class 'as a discursive rather than an ontological reality' (8). Social historians of the Stalinist era have generally opted for a softer form of this analysis: they have emphasised the importance of languages of class in the formation of class consciousness, but have shrunk from explicitly denying the objective existence of class as a social relation.[12] Commentators such as Hoffman (1994), Siegelbaum and Suny (1994), Kotkin (1995) and Smith (1997) have nevertheless stressed the key role of language in the co-option of Soviet workers, arguing that because the regime monopolised the language of class, workers were unable to define their own interests in class terms, and were thus prevented from forging effective opposition to the regime:

> The language of class was appropriated by party and state managers, and though autonomous discourses of workers existed within the totalizing dominance of the regime, workers had no formal modalities for wide expression and reproduction except through the state. (Siegelbaum and Suny, 1994: 26)

> Soviet industrialisation provided no clear class enemy and party officials prevented the voicing of any anti-establishment ideology that might have created a sense of common interests and class solidarity among workers. Not only did the authorities maintain tight control over the media and other ideological resources, they appropriated the language of class and of revolution to serve as the language of the Soviet establishment. Even if workers did not believe that the Soviet Union was a working class state they could not use the language of class to stir rebellion when Party leaders had co-opted it as a means of political legitimation. (Hoffman, 1994: 211)

This approach to explaining the apparent[13] quiescence of workers is superficially persuasive. If Soviet workers were not merely repressed, but prisoners of language, unable to articulate dissent within the terms of the regime's hermetic and all-pervasive discourse, then it is not surprising that, like Orwell's subjects of 1984, they showed little sign of resistance. But this approach actu-

[12] Stedman Jones himself is ambiguous with regard to this question, but there are certainly points at which he argues against the existence of a social world prior to and outside of its articulation in language (Kirk, 1994: 226).

[13] The level of protest in the Stalin era may have been underestimated. Already it is clear that the regime faced serious resistance in the early phase of industrialisation, which forced some modification of policies. An important example of this is provided by Jeffrey Rossman in his account of the wave of strikes involving 16,000 textile workers in the Ivanovo Industrial Region in 1932 (Rossman, 1997).

ally raises more questions than it answers. First, as both the above quotations reveal, it is not clear how much explanatory significance proponents of the discursive approach are prepared to assign to the regime's linguistic monopoly. Siegelbaum and Suny slip in a reference to the fact that workers depended on the state for their 'reproduction', while Hoffman stresses that Soviet industrialisation 'provided no clear class enemy': neither is prepared to deny the existence of 'objective' barriers to workers' resistance. This ambiguity immediately raises the question: was the main problem the fact that workers lacked the language with which to *create* a sense of common interests, or were there actually objective reasons why workers could not *perceive* a 'clear class enemy' and mount effective resistance within the framework of the Soviet system? This question is particularly pertinent given that Rossman (1997) and Davies (1997) have recently argued, on the basis of their analyses of conflict and discontent in the 1930s, that the Party's language of class was not a barrier to the expression of discontent, but, on the contrary, could act as a 'double edged sword' (Rossman, 1997: 9), which could be employed by workers to defend their interests.

Searching for the 'class enemy': class contradictions in Soviet Russia

What the contribution of the discursive school serves to highlight is that 'languages of class' cannot be analysed in isolation from concrete social relations. Without a distinction between social reality and discourse it becomes impossible to understand the origins of particular 'discourses'. The means through which class is perceived and articulated are complicated. But, as the analysis of workers' collective identification to be developed in this book illustrates, workers' subjectivity is certainly related to the social relations of production and the social organisation of work. So what was it about Soviet social reality that made it difficult for workers to define and defend their interests?

All the protagonists in the debate regarding the perplexing quiescence of Soviet workers implicitly agree on one thing: Soviet workers had objective reasons to resist. The question is, who or what were workers to oppose? The Soviet 'ruling class' was notoriously elusive, as Alec Nove's famous article on the subject illustrates (Nove, 1989). The central problem in defining class relations under Soviet Communism is that there was no private ownership of the means of production, so that everyone, up to and including the General Secretary of the Party, was formally a wage labourer. Nor was social production subordinated to the law of value: Soviet enterprises did not need to make a profit, but only to make the centrally imposed plan, and Soviet goods were therefore produced as physical use values rather than exchange values. Thus, Soviet social relations were not constituted on the basis of the opposition between a propertyless class, whose members were compelled to sell their labour power, and a propertied class of capitalists: the defining division of capitalism – that between capital and labour – was absent. Nevertheless, workers were required to work in order to secure their means of subsistence, and they had no control over what they produced: as Bob Arnot argues, the ruling group attempted 'to extract and control the distribution of a reliable, useful and growing surplus from the direct producers' (Arnot, 1988: 30).[14] This surplus extraction was required to reproduce the regime's position both domestically (through control over distribution of benefits and privileges) and internationally (by building up its military capability). Thus, the Soviet system can be characterised as one of politicised surplus extraction, in which the central antagonism lay between the direct producers and the economic administrators of the state. Within this system, position within the political hierarchy was the main determinant of status and access to resources.

Does this imply that Soviet society was divided into two classes? This is a more complicated question. First, the position of the 'ruling group' is hard to define, given the fact that they did not actually own the property over which they exercised control. Simon Clarke, for example, whose characterisation of the Soviet mode of surplus extraction is similar to that put forward by Arnot, leaves open the question of whether the 'ruling stratum' constituted a class (Clarke *et al.*, 1993: 7). Meanwhile, the identity of the 'direct producers' is also confused. For the entity from which the surplus was extracted was the enterprise rather than its workers. That is, the workers and managers of a given enterprise actually had a common interest, in relation to the state planning agencies, of maximising inputs and minimising the level of extraction (through overestimating the resources needed to fulfil the plan and underestimating the productive capacity of the enterprise). In this way, the volume of work was informally regulated and everyone could hope for a bonus when the plan was fulfilled: this is what Clarke refers to as the informal 'production pact' between Soviet managers and workers (Clarke *et al.*, 1993: 99). This analysis implies, however, that workers cannot be unambiguously identified as a class of direct producers from whom the surplus was expropriated. For, as Arnot emphasises, the role of Soviet enterprise managers was highly contradictory. Although in one sense they made up part of the political elite it was also rational for them, as well as for workers, to take action which contradicted the aims of their political masters (Arnot, 1988: 40): they too were in conflict with the expropriating 'centre'.

The common interests of workers and managers found their expression in the ideological and social form of the labour collective (*trudovoi kollektiv*), which was the Soviet name for the social community of the enterprise which included all employees, from the director to the cleaner. As Vladimir Ilyin argues, it was the labour collective which was the subject of the primary contradiction in Soviet society, which lay between the enterprise and the external administrative environment (Ilyin, 1996a: 38–9). But, as will be seen below, the common

[14] As Arnot asserts, the surplus should be defined as the whole socially produced surplus over and above what is necessary to reproduce the direct producers (Arnot, 1988: 30).

interests of workers and managers in relation to the outside world co-existed with sharp conflicts of interest between the two groups *within* the enterprise. It is this contradictory relationship between workers and their managers which constitutes the main problem of defining the class interests of Soviet workers.[15] The labour collective was simultaneously a genuine community of interest and a site of struggle; for this reason it really was the case that workers faced an elusive enemy. As this book will show, this contradiction had a huge influence on workers' subjective collective identification, both in the Soviet era and during transition.

The labour collective

One of the main questions which arises from the discussion of the previous two sections is, how have commentators arrived at such divergent views of Soviet workers? Were workers disaffected and 'atomised'? Were they prevented from expressing discontent through the regime's monopolisation of potentially subversive discourses? Or were they voluntarily compliant and incorporated through positive means? The key to reconciling these divergent views lies in understanding the contradictory social form of the labour collective.

The labour collective was both a real and a symbolic community. The definition of the enterprise workforce as a collective had an important ideological significance for the Soviet authorities. The idea of the *kollektiv* was part of Bolshevik culture in the pre-revolutionary period, and was associated at this time with the 'God-building' tendency within the Party. Although Lenin labelled God-building a deviation, the notion of collectivism retained the spiritual dimension attributed to it in the writing of Anatoly Lunarcharsky, Aleksandr Bogdanov and Maxim Gorky (Kharkhordin, 1996: 4–6). After the revolution the name *kollektiv* was reserved for groups closely connected to the communist cause, and in most cases referred to a Party cell. Then, in the 1930s, Stalin's emphasis on individual responsibility – which reached its zenith in the Stakhanovite campaign – meant that collectivism became somewhat suspect, while the link with the Communist Conscience which had existed in everyday usage also weakened (Kharkhordin, 1996: 10). By the 1950s, however, the term *kollektiv* began to be used to refer to almost all groups within Soviet society: 'Collectives were everywhere, one entered a collective as a small child, passed from one to another in the course of one's life, but one was never (normally) outside a collective. The network of collectives constituted the entire terrain of social life' (Khakhordin, 1996: 12). The generalisation of the term *kollektiv* meant that it lost its specific political connotation, but Soviet social science of the 1970s still defined a *kollektiv* as a group which served socially defined goals

[15] This contradiction between the enterprise and the outside world is also felt in a different form under capitalism. In troubled times it is common for trade unions to restrict their demands in the interest of the survival of particular workplaces: indeed, 'concession bargaining' is now rife in the industries which face competitive product markets.

rather than a narrow group interest. To pursue the latter was to be a 'false collective'.

The enterprise workforce was thus a *kollektiv par excellence*, serving the cause of communism by meeting the plan. The communist emphasis on industrial production and the working class meant that a labour collective was a particularly hallowed form of *kollektiv*; it was a concept with spiritual, moral and mythical dimensions, born of the heady combination of the communist reverence for collectives and sense of the dignity of labour. The importance of the labour collective was highlighted in a variety of symbolic ways. For example, outside most enterprises would be an honour board displaying photos of workers who had distinguished themselves in some way, while traditions such as the provision of presents at New Year for workers' children emphasised the idea that the enterprise was as much a community as a workplace. Workers also had a responsibility towards the collective which was underlined by their (compulsory) participation in the *Subbotnik* – an occasional unpaid working Saturday. Meanwhile, individual work groups within the enterprise – also referred to as collectives – had both a symbolic and a real influence over the life of the individual worker. For example, the 'opinion of the collective' would be taken into account in court if a worker was charged with a minor offence, and a judge could place a defendant who had been found guilty under the supervision of the work collective. Although in such situations the deliberative role of the work collectives concerned was largely formal – the view of managers or Party officials was decisive – the symbolic significance accorded to the 'opinion of the collective' underscored the fact that Soviet enterprise was conceived as a moral as well as a productive community.

Along with the stress on collectivism, the regime also propagated the idea that work (especially of an unpleasant, physical variety) was inherently meaningful and noble. Enterprises were adorned with statues, pictures and murals of heroic workers emblazoned with slogans such as 'Glory to the miners!', while 'days' were nominated to honour different professions, so that the trade union of a large enterprise containing a myriad of different types of worker would be forever organising festivities.[16] The fact that work itself was accorded a moral significance only heightened the status of the labour collective as a key locus of meaning in Soviet society.

The labour collective was more than an ideological nomenclature, however. It was the concrete expression of the common interests of its members, and constituted the basic unit of Soviet society. State protection and benefits were mediated through the labour collective, so that the majority of workers' needs – from the provision of kindergarten places to payment of funeral expenses – were met within the collective. Moreover, in the case of the

[16] A mine, for example, would not only celebrate the 'day of the miners', it would also mark the 'day of the medical workers', the 'day of trade union workers' and so on. Often parties would be held at enterprise leisure facilities for the workers concerned.

most important forms of provision there was often no alternative means of securing access to the required goods and services other than through the workplace. The average large enterprise provided housing; rest and leisure facilities, such as 'palaces of culture', sanatoria, holiday resorts and pioneer camps for children; kindergartens (which, since Russian children begin school at the age of seven, also provide the first years of schooling); prophylactic health care facilities, and sometimes local primary health care facilities; allotments; access to a variety of goods in short supply; and grants, soft loans and guarantees for loans from local retail establishments.

Workers' relationship to the labour collective was not only defined by need, it was also characterised by genuine attachment. The enterprise was a social focus. The corollary of the provision of social and cultural facilities through the workplace was the impoverishment of the material conditions of private existence. Cramped housing conditions and the lack of leisure facilities meant that work was an escape from the drudgery of home and a welcome opportunity for communal sociability. As Alasheev has argued, workers 'live in such conditions that work . . . is the single socially approved possibility of self realisation' (Alasheev, 1995b: 71). This is strongly reflected in the way in which workers talk about their enterprises. Perhaps surprisingly, as will be seen in later chapters, interviews with workers reveal that, though they may not have accorded 'the collective' precisely the same significance as the authorities did, they too felt attached to the labour collective and their immediate work collectives. For example, the sense of common endeavour fostered by attempts to meet the plan is often referred to by workers, as is the importance of their workplace as a community, or 'second family'.

The integration of Soviet workers

The preceding account would seem to add weight to David Lane's claim that the Soviet working class was 'incorporated'. But it is not sufficient to see workers as integrated into cosy collectives; the contradictory social form of the labour collective was reflected in the mode of workers' integration within it.[17] These contradictions are revealed most starkly in the system of social provision within the enterprise.

The system of provision can be characterised as one of state paternalism mediated through the labour collective. This provision had two key features. First, it was monopolistic in character: most of the goods and services available through the enterprise could not be acquired readily on the open market.

[17] One of the ways in which the argument advanced here differs from most earlier interpretations of workers' relationship to the Soviet state is that it sees the integration of workers as something which took place within the enterprise, rather than at some abstract societal level. Andrew Walder similarly sees relations within the enterprise as the key to the political control of Chinese workers, although in the Chinese case the Communist Party played a far more significant role within the enterprise than did its Russian counterpart (Walder, 1986).

Within the Soviet system money did not, as under capitalism, regulate the production and allocation of goods and services (including labour power), and thus it did not act as a universal equivalent.[18] Instead, production and allocation were regulated administratively. Many goods could not be paid for in roubles: in the Soviet Union it was not just love that money could not buy, but a wide variety of concrete objects. The state retail sector was poorly developed and workers were often unable to spend their income. As shortages grew worse in the Brezhnev era, the level of savings grew. By 1980 the total savings in the Soviet Union were equal to seven months of national retail (state and co-operative) trade turnover (Connor, 1991: 119). The fact that people were able to save so much from what were low wages clearly indicates the marginal role of money within the system. In this environment, *access* to goods in short supply was more important than the means to pay for them. The goods provided by enterprises were therefore not perks but necessities: survival outside a labour collective was almost impossible.

Second, the distribution of benefits within the enterprise was discretionary. The Soviet system of administrative allocation resulted in an economy which was plagued by shortages,[19] and enterprise social provision in particular was always characterised by its inadequacy. For example, nearly all enterprises had long housing waiting lists and there was usually competition for kindergarten places, holiday vouchers and so on. The individual managers and members of the trade union committee who administered these benefits thus possessed a potent means of control: workers were compelled to cultivate individual relationships with authority figures in order to secure access to goods and services which were in short supply. One of the central problems of paternalism identified by the sociologist Howard Newby is that while it tends to disguise fundamental conflicts of interest, it also 'grants subordinate individuals certain prerogatives which . . . tend over time to be appropriated as "rights"'. But, as Newby points out, this contradiction can be managed through face-to-face contact (1977: 70–1). This is exactly how it was managed in the Soviet context: the collectivist potential that the sociologists Abercrombie and Hill (1976: 414)[20] discern in paternalism was negated because Soviet state paternalism

[18] The idea that money is defined by its form as the universal equivalent was first advanced by Marx in *Capital* (Marx, [1867] 1990: 180–1). Ticktin (1992) also uses this definition of money to argue that the Soviet Union was a non-monetary society.

[19] This was one of the manifestations of the conflict of interest between labour collectives and the 'centre': enterprises had no interest in maximising productivity. In the case of consumer goods and services, shortages were exacerbated by the political (and hence allocative) priority given to heavy industry.

[20] Abercrombie and Hill's definition of paternalism is based on the idea that it is 'a collective form of social organisation' in which 'all subordinates stand in the same relation to the paternalist' (1976: 414). This is contrasted with patronage, which is seen as an individualistic form of relations, based on face-to-face contact. They moreover argue that 'paternalism is a relation which entails "total involvement", while patronage is partial' (416). The Russian example does not fit this clear-cut distinction. First, enterprise paternalism takes the form of patronage of individual

took the form of patronage of individual workers by individual managers within the enterprise. Moreover, the divisive effect of this discretionary distributive regime was reinforced by the informal and individualised form of enterprise punitive regimes and payment systems.[21]

This feature of the social relations of the labour collective explains why Soviet workers have been seen as both 'incorporated' and 'atomised'. Workers were bound to the system through what Lane calls the 'positive forms of incorporation' (1985: 163) such as social provision within the labour collective, but they were also divided and encouraged to respond to the administration as individuals, as the supporters of the 'atomisation' thesis claim. Soviet workers were constituted as members of a collective, but only 'communist' collectivism was encouraged. Any forms of 'false' collectivism were discouraged by the individual dependence of workers on their superiors, which meant that workers' instinctive reaction to a problem at work was to attempt to 'fix' it using personal contacts or by applying direct to a member of management.

It is important to emphasise that the argument that only individualised means of resolving problems were open to workers is not the same as that which claims they were 'atomised'. As Andrew Walder has argued with regard to China, rather than being atomised, workers under communism were involved in a 'rich sub-culture of instrumental-personal ties independent of the party's control' (Walder, 1986: 7). Such involvement was necessary to obtain official approvals, housing and other public and private goods controlled by low-level officials, and though such ties were not officially approved of, they were structurally encouraged by a system which created scarcity and left many questions of distribution to the discretion of enterprise and local officials. On one level, the effect may have been the same as that posited by the 'atomisation' thesis: collective organisation proved almost impossible. But, looked at in terms of the potential for change, the two ways of viewing the working class have quite different implications. 'Atomisation' imposed from above by repression would only last as long as the legacy of fear, whereas networks of profitable personal ties developed to circumvent the limitations imposed by a bureaucratic state would not merely dissipate as the state began to change.[22]

workers by individual members of the administration, and second, it could be argued that the 'total involvement' of Soviet workers in their enterprises was defined as much by their dependence on the patronage networks at work as by the construction of the paternalist enterprise as the centre of their world.

[21] On punitive regimes see Alasheev, 1995b: 52–8; on payment systems, Vedeneeva, 1995: 224–39, and Donova, 1996: 41–62.

[22] Abercrombie and Hill argue that, while paternalism can adapt to 'become compatible with the central principles of modern society', as they claim it has in Japan, patronage is in opposition to these and 'continues so long as modern institutions fail to regulate social life according to legal-rational criteria' (1976: 416). This formulation implies that there is one form of 'modern society' – one governed by legal-rational criteria – towards which 'less developed' nations are progressing. According to such a view, the marketisation of Russia should begin to sweep away patronage networks. But within the Soviet enterprise it was patronage rather than paternalism (understood as a

Meanwhile, the idea of 'incorporation' implies a level of consensus which did not exist within the Soviet enterprise. Compared to Japanese workers in large firms operating a policy which has been termed 'welfare corporatism' (Dore, 1973: 275), for example, Soviet workers were outrageously insubordinate. Indeed, if the attitudes and behaviour of Soviet workers are measured against those described in Ronald Dore's classic comparison 'British Factory – Japanese Factory', they are much closer to those of the British workers.[23] And yet the resistance of Soviet workers was contained and never attained an organisational expression as it has in Britain: they may have outdone the British in cursing and absenteeism, but collectively Soviet workers were far weaker than their Japanese counterparts. This underlines the fact that the integration of Soviet workers was a complex process: hostility to authority co-existed with identification with the enterprise; conflict was endlessly created and contained. As will be shown in later chapters, a whole series of such contradictions was engendered by the fact that the common interest of the labour collective co-existed with an objective conflict of interests between the workers and managers within its ranks (Ilyin, 1996a: 35–40).

What does this imply about the form of collectivism fostered within the labour collective? The collectivity of the labour collective can be defined as alienated, in the sense that workers related to the collective as something standing outside of and opposed to themselves: it was neither perceived nor realised as the product of the collective organisation of individual workers or the association of their immediate work collectives. The embodiment of the collective was the enterprise director, who was expected to represent and defend it in relation to the outside world. The labour collective was an entity to which workers appealed and from which they received: workers identified with the labour collective but as supplicants rather than subjects; the labour collective was their guarantee of security but it was also the site of their subordination.

It was therefore not the scale of workers' dependence on the labour collective which explains their integration, but its form. Stephen Crowley (1997) has recently attempted to analyse workers' integration within the Soviet system in terms of their relation to the enterprise, but he places all the explanatory significance on the level of benefits they received. Thus, he argues that miners have been more militant than steel workers during the transition era because they receive fewer non-monetary benefits and therefore have less to lose

collective form of social organisation) which was the key to managerial power. This calls into question Abercrombie and Hill's view of the durability of the respective systems, as does the experience of 'modern' countries, such as Italy, where patronage networks have to date proved ineradicable.

[23] Dore encapsulates the resentment of British workers with a long excerpt from an interview with a shop steward at the firm he calls 'English Electric'. This worker's critique of 'bloody disorganised' management could easily have been culled from an interview with a Russian worker. The key beliefs that Dore singles out as characterising the British workers' attitude to authority – that managers are 'overprivileged . . . inefficient . . . and . . . owe their jobs to influence rather than merit' – are also constantly encountered among Russian workers (Dore, 1973: 251–3).

through taking action (195). But, as one of the reviews of the book pointed out, 'on this argument women workers would be more militant than men, workers in light industry than miners, in small enterprises than in large enterprises and surface mineworkers more than faceworkers' (Clarke, 1998: 722). In contrast, this book will argue that what has structured workers' response to transition is not their dependence *per se*, but the alienated form of collectivism which characterises workers' relationship to the labour collective and defines their individual dependence.

The specificity of Soviet state paternalism

The idea that workplace paternalism can be employed as a means of avoiding conflict and fostering identification with the employer is not new, and it is a management strategy employed across a range of different societies. Was there anything distinctive about Soviet state paternalism? On Crowley's argument the answer is no, and the quiescence of Soviet and post-Soviet workers can thus be reduced to a matter of individual instrumental rationality indistinguishable from that of any other group of workers with something to lose in the event of a conflict with their employer. But Soviet state paternalism did differ from forms of workplace paternalism which exist under capitalism. Under capitalism, paternalism is a particular managerial style which firms can choose or not according to its perceived profitability. By contrast, Soviet state paternalism was an integral part of the Soviet industrial system, or, as the Samara Research Group put it, Soviet paternalism was not 'a particular kind of management strategy' but a 'particular form of production relations' (Samara Research Group, 1995: 128–9).[24] This meant that Soviet enterprise social provision had a coercive character which distinguishes it from forms of capitalist enterprise paternalism. For example, enterprise paternalism is a common form of industrial organisation among large enterprises in Japan, which usually provide a wide array of benefits for the permanent part of their workforces, such as housing, health insurance, recreational facilities and pensions. But although workers in Japan are provided with benefits, they have alternatives; they are not forced to depend on their firms. Consumer markets in Japan are fully developed and are not characterised by shortages: there is, for example, a free market in housing, even if the terms on which it can be obtained through the workplace are more attractive (Walder, 1986: 23).[25] The crucial difference between the

[24] The Samara Research Group identifies the main features of Soviet enterprise paternalism as being the charismatic leadership of the director, who is seen as the 'father' of the collective; strongly hierarchical management; closed information, which supports the authority of the director; egalitarian principles of wages and distribution, which, though not always practised, are expected; non-monetary relations; and the ideology of paternalism, which the Samara Research Group sums up as 'a particular atmosphere, a sense of collectivism, unity, of a common business, supported by the ideology of socialist society' (Samara Research Group, 1995: 137).

[25] Nevertheless, the dependence of Japanese employees is substantial. Employees in the elite

position of workers within the two systems is that the wages of one group can purchase most of the necessities of life, even if at a pinch, while Soviet workers could not purchase a variety of goods and services with the money they earned, but had to apply to their enterprises to obtain access to them.[26]

In terms of the internal administration of social provision, it is again the element of compulsion in the Soviet system of paternalism which chiefly distinguishes it from its capitalist counterparts. According to Dore's characterisation of Japanese enterprise paternalism, the mode of internal administration is quite different from that of the Soviet enterprise. In Japan provision is adequate and the mode of distribution is not designed to foster dependency: benefits do not have to be competed for, begged for, bribed for:

> The favours that count materially – the welfare benefits, the housing, the educational loans and the dormitory accommodation for one's children studying in Tokyo – are all specified and contractual: their distribution rule-bound and institutionalised as employee 'rights'. And the loyalty which is bought by these favours is an institutional, not a personal loyalty. . . . The personal paternalism of the foreman and section chief . . . could disappear entirely without having any effect on the main institutions of 'welfare corporatism'. (Dore, 1973: 274–5)

By contrast, there was nothing 'contractual' or 'rule-bound' about Soviet paternalism: it was personal, particularistic, discretionary. As has been argued, the personal dependency of workers on their managers was a crucial means of integrating Soviet workers. However, the uniqueness of this phenomenon in itself should not be exaggerated. Dore paints a very benign portrait of life in the Japanese enterprise and many commentators have disagreed with his account. Although the distribution of benefits may not be discretionary in Japanese enterprises, other commentators have argued that subjective merit ratings and personnel evaluations play a vital role in fostering Japanese workers' apparent devotion to their jobs. That is, if Japanese workers do not unconditionally submit themselves to the will of their managers, attend quality control circles in their free time and consistently go beyond the call of duty, they risk missing

20 per cent of large firms receive substantially better pay and benefits than workers in small firms (Chalmers, 1989: 49) and they have big incentives to stay put. Within the elite sector of the economy, career progression is expected to occur within internal labour markets (Kumazawa and Yamada, 1989: 108), and the system of paternalism reflects this. For example, in contrast to the situation in Soviet Russia, Japanese employees who leave their enterprises are not entitled to keep any accommodation they have acquired from the firm.

[26] It should be noted that the extent of Japanese provision is far greater than that of most 'paternalist' firms in the West. For example, Martin and Fryer, in what is intended to be a case study of capitalist paternalism, cite very few tangible benefits provided to the workers at 'Casterton Mills', the firm in Lancaster supposed to exemplify this form of organisation (Martin and Fryer, 1973). It is apparent from the account of Martin and Fryer that the 'paternalism' they describe is largely a matter of style rather than substance. Later commentators such as Warde (1989), however, have argued that even this circumspect account exaggerates the extent of paternalist practice in Lancaster.

out on the merit-related pay increases, promotions and benefits which over time have a significant impact on pay (Kumazawa, 1996: 251–2). The discretionary nature of the Soviet system was therefore not in itself distinctive – subjective evaluation is a common means of controlling workers.[27] What distinguished it was its coercive form: Soviet workers had no alternative but to cultivate personal ties in order to survive. As will be seen, this pervasive personalisation of relations, combined with the alienated collectivism fostered within the labour collective, has had a crucial influence on workers' behaviour during transition.

Trade unions under communism

One side of the explanation of workers' quiescence during transition lies in the form of their integration under communism, which, as later chapters will argue, has structured their behaviour during the transition era. The other side of the question concerns the role of the former communist trade unions during transition. These trade unions organised the overwhelming majority of Soviet workers in the past, had a primary organisation in every enterprise, and had vast resources at their disposal. Under communism, their subordination to the Party denied them an independent voice, but why have they been unable to transform themselves into independent organisations representing workers' interests during the transition period? Trade union leaders at every level have expressed a desire to change, to become 'real trade unionists', so why do surveys still show that trade unions are neither trusted nor respected by their members? Why have the unions failed so dramatically to defend their members in the face of the catastrophic decline of their living standards? The main argument pursued in this book is that this is not primarily a personal or a strategic failure, but is a structural problem, related to the form of trade unions' insertion within the communist system. The following sections of this chapter therefore explain the role and position of the trade unions in the past in order to provide the basis for the analysis of their response to transition.

The theoretical basis of Soviet trade unionism

The famous slogan of the early period of Stalinist industrialisation instructed union officials to turn their 'faces towards production' (Ruble, 1981: 13), meaning that they were to direct their attentions to the overriding task of plan fulfilment. The unions were deprived of any independence that they had retained during the period of the New Economic Policy (NEP): those who

[27] Most perfomance-related pay systems other than piecework contain a subjective element, and like the discretionary elements of the Soviet system this is often perceived as unjust. See, for example, Marsden and Richardson, 1994.

opposed the new policy were dismissed as 'right deviationists'. The Soviet model of trade unionism was not simply a product of the Stalin era, however. It was during this era that the unions settled into their specific role and position within the Soviet polity and enterprise, but the theoretical basis of Soviet trade unionism had already been laid down by Lenin.

Lenin's position partly emerged from the need to reach a compromise between the contending factions at the Tenth Party Congress in March 1921, but it also exhibited familiar Leninist preoccupations, such as the vanguard role of the Party and the need for rapid industrial growth to combat Russia's long-lamented backwardness. Lenin negotiated a middle path between the syndicalist programme of the Workers' Opposition, which sought to give unions control over industrial production, and the 'production' platform of Leon Trotsky and Nikolai Bukharin, which advocated the statisation of the trade unions, and their transformation into 'production unions' that would manage compulsory labour programmes, improve productivity and enforce labour discipline. The victorious proposal of the Group of Ten supported by Lenin allowed for a circumscribed form of union independence within the boundaries determined by Party policy.[28] This compromise between Party factions formed the basis for Lenin's prescription for 'The Role and Functions of Trade Unions Under the New Economic Policy', which was adopted as policy by the Central Committee on 12 January 1922, and was published five days later in *Pravda* (Carr, 1952: 326). The arguments Lenin developed in some respects referred specifically to the NEP period in which the market and private trading were restored, small-scale private enterprise was permitted, and 'large-scale' (factory) production was reorganised on commercial lines. But Lenin's view of the proper relationship between workers, trade unions and the Soviet state was not only applicable to this period and came to form the theoretical justification of the role of trade unions in the Soviet state thereafter.

The commercialisation of production in the NEP period would, according to Lenin, 'inevitably give the masses the impression that there is an antagonism of interest between the management of the different enterprises and the workers employed in them' (Lenin, [1922] 1947: 760), and in this situation the job of the trade unions was to defend workers' interests and to try to ensure that their living standards were raised. To this end the trade unions were permitted to 'constantly correct the blunders and excesses of the business organisations resulting from the bureaucratic distortions of the state apparatus' (761), which would in turn further the cause of socialism because 'the ultimate object of every action taken by the working class can only be to fortify the proletarian state and the proletarian class state power' (762). This last sentiment gets to the crux of the Leninist position on Soviet trade unions: since under socialism, the interests of the (socialist) state and the working class were deemed to be identical, the role of unions was reduced to one of 'mediators' (762), permitted to

[28] For more details on this debate see Carr, 1952: 222–7.

avert the disputes arising from 'distortions' through a 'foresighted policy' (763). Disputes over union priorities were to be settled by a 'higher authority': the Communist Party (767), which alone comprehended where the objective interest of the state and working class lay.

The common interests of the workers and the socialist state (run by the vanguard of the working class, the Communist Party) might not be immediately apparent to the workers, however. This possibility did not elude Lenin, who was quite clear about the fact that trade unions should be under the control of the Party. Trade union groups should be run by 'responsible comrades', though not necessarily Party members (766),[29] who should be close to the workers and able to 'stoop to their level', but at the same time able to resist the temptation to 'pander to the prejudices and backwardness of the masses' (767). Thus, in place of self-organisation among workers, the 'responsible comrades' were to 'judge the mood, the real aspirations, needs and thoughts of the masses' (766). Lenin recognised that one of the greatest dangers facing the Communist Party was 'divorcement from the masses' – a grave risk considering the self-appointed vanguard of the working class had no reliable institutional mechanisms for maintaining links with the class in whose name they exercised their dictatorship – and the trade unions were supposed to avert this danger by becoming, in Lenin's famous phrase, 'transmission belts from the Communist Party to the masses' (766).

Although Party leaders wanted to be kept informed of the workers' mood, however, they believed themselves to be fully cognisant of the 'real' interest of the working class. For Lenin and all the Bolsheviks this was clear cut: it lay in increasing the output of manufactured goods by 'enormous dimensions' (763). The trade unions were therefore charged with enlisting 'the working class and the masses of working people generally for all branches of the work of building up the state economy' (765). They should explain to workers that this was in their interest, while fighting to improve labour discipline. In addition to this, the trade unions were to act as 'schools of communism' (764), training the future managers of Soviet industry and providing a channel for the advancement of workers.

The later position of the trade unions under state socialism was not definitively anticipated by Lenin and the role of the unions was progressively refined, most notably through the addition of social and welfare functions within the enterprise. Aside from the social duties that the unions acquired, however, the main features of Soviet trade unions can be traced back to the Leninist ideological heritage. First, the unions were never intended to be an

[29] The Eleventh Party Congress in 1922 decided that only Party members of several years' standing could be elected to leading posts in the trade unions. As E. H. Carr remarks with reference to this decision, 'The fate of the trade unions was an excellent illustration of the way in which the NEP, by conceding a measure of economic freedom, provoked a strengthening of direct political control by the Party over individuals or organs which might be tempted to abuse this conditional freedom' (Carr, 1952: 328).

expression of workers' self-organisation: they were to be subordinated to the control of the Communist Party, 'which alone will be capable of withstanding the inevitable petty-bourgeois vacillations . . . and the inevitable traditions and relapses of narrow-craft unionism or craft prejudices among the proletariat' (Lenin, [1921] 1947: 684). Second, the supposed coincidence of the interests of the socialist state and the working class meant that the unions' role was minimised to that of mediators. And though the unions were formally allowed to correct the 'distortions' of the bureaucratic state, the definition of what was and was not a distortion was ultimately in the hands of the 'higher authority'. Thus, the defensive function of unions was even in theory whittled away to tinkering within the parameters set by the state. Lastly, Lenin laid the foundation for the co-operation between the unions and management by asserting that part of the unions' role was to press for improvements in productivity and labour discipline, since this was in the objective interests of the working class. Lenin did, it should be stressed, genuinely struggle with the problem of the relationship between the socialist state and the working class, but this was not the case with the later Soviet theorists who dealt with the question of trade unions. They trotted out a neat formula whenever it was required, in unreflective justification of trade union subordination: there was no conflict of interest between the Soviet state and working class, and trade unions existed to deal with the non-class economic struggle against bureaucratic distortions, themselves conveniently passed off as survivals of capitalism.[30]

After Lenin: the fate of the unions

Under Stalin, the trade unions lost any defensive functions they had retained during the NEP period and were involved in the quest for increased production. The unions' chief concerns became those of 'socialist emulation' and other production-raising strategies. From 1929 to 1934 the central union agencies even attempted to use collective agreements as an instrument for increasing productivity (Ruble, 1981: 17), but this attempt proved ineffective and collective agreements ceased to be concluded altogether between 1937 and 1947.

Meanwhile, the unions also acquired a role in welfare provision within enterprises.[31] After the abolition of the People's Commissariat of Labour in 1933, the unions accepted responsibility for the distribution of welfare benefits.[32] Given the tightly circumscribed representational role of unions and

[30] See, for example, Kuusinen, 1961: 529.

[31] An earlier, short-lived attempt to put social insurance under the auspices of the newly created All-Union Central Council of Trade Unions (VTsSPS) had been made between 1917 and 1921 (Madison, 1979: 85).

[32] These included sickness and invalidity benefits, maternity benefits and child care allowances for low income families, but not unemployment benefit, which was abolished in October 1930. These benefits, though distributed by the unions, were financed through state tax of 4.4–14 per cent against each ministry's and enterprise's pay roll.

their limited effectiveness in the struggle for increased productivity, social administration quickly became their most important practical function within the enterprise. It filled the days of union officers and defined their position as part of the enterprise administration: effectively the unions functioned as enterprise social welfare departments. In this capacity the unions gave priority to the interests of management rather than workers – in 1938 factory trade union and management officials were required to consider attendance and production records in the distribution of welfare benefits (Ruble, 1981: 24). Although this requirement was annulled in 1951, in practice the control of welfare benefits gave the unions a good deal of discretionary power over workers, and they were expected to use it: there were, for example, eighteen different categories of disability allowance, which could be granted or not according to criteria such as length of service, work record and the role played by alcohol in the disability (88).

The shift from coercion to encouragement as a means of promoting industrial growth which characterised the Khrushchev era obviously had implications for the role of the unions. Their role was enhanced by a resolution of the Central Committee of the Communist Party in December 1957. This demanded increased worker involvement in production administration and also accepted the need to improve the education, safety and welfare of Soviet workers. The resolution stated too that the collective agreement was the juridical foundation for all trade union activity at the enterprise level. Nevertheless, promoting production remained the unions' primary obligation, and their relationship with the Communist Party was not altered by the resolution (Ruble, 1981: 33–4). The 'central task' of the unions, defined in the preamble to their by-laws, was unambiguous and unchanged from that defined by Lenin: 'to mobilise the masses for the attainment of our principal economic goal – the creation of the material and technical basis of communism, for the further strengthening of the Soviet Union's economic and defence power, for ensuring a steady rise in the people's material and cultural standards' (Godson, 1981: 113).

The functions of Soviet trade unions

The traditional arguments about Soviet trade unions focus on the issue of whether the trade unions had the 'dual function' of imposing the policies of the Party from above at the same time as defending workers' interests (Ruble, 1981) or whether they were simply part of the repressive apparatus (Conquest, 1967; Schapiro and Godson, 1981). Clarke *et al.* (1993: 93–4) have challenged these arguments, however, claiming that at the enterprise level the trade unions were an instrument of the enterprise administration and that this was in fact their dominant role. Their 'dual functions' therefore involved supporting enterprise management and acting as a 'transmission belt' for the Party. From the Khrushchev period, along with raising productivity the unions were supposed to support the social policy of the Party within the enterprise, by defending individual workers against unscrupulous management. Thus the unions were partly constituted as a check on management, and could receive support from the Party in this capacity. The extent to which they could perform this role on an everyday basis, however, was limited by their close integration with management: serious confrontations between enterprise management and unions would only occur if the union was prompted and backed by the Party.

The main characteristics of the official trade unions were, first, that they organised approximately 98 per cent of the workers. The unions were organised on the branch principle, grouping together all the employees (including the managers) from a particular branch regardless of their function, and belonging to a central confederation, the VTsSPS.[33] The primary enterprise trade unions were grouped together in regional organisations as well as branch organisations. The trade unions were governed by the principle of 'democratic centralism', which in theory combined centralised command with the initiative of the rank and file, though in practice there was no space for the latter 'democratic' element. There were supposedly elections for all posts in the union, but these also were governed by the principles of democratic centralism – those voting took part in a 'democratic' procedure constrained by the strongest centralism, in which the candidates were nominated from above.

A formal statement of the functions of Soviet trade unions by an official of the USSR State Committee on Labour and Wages (Piatakov, 1962), published for consumption in the West, listed the following functions of enterprise trade unions: participation in the drafting of production and investment plans; the receipt of reports from management about production plans; expressing an opinion on candidates for management, and monitoring management performance on labour issues (the lax could be reported to the 'responsible bodies'); participation in decisions regarding job evaluation; and the organisation of production conferences. The functions of 'examination of labour disputes' and monitoring health and safety appeared near the bottom of this list, though it did also include higher up the conclusion of collective agreements with management, and the representation of workers over issues of work, living conditions and culture. The trade unions also, as mentioned above, administered social insurance and social facilities provided at an enterprise level such as sanatoria, holiday facilities and so on.

Although such a range of duties might not seem particularly exceptional, precisely those issues with which the Western observer would expect the trade union to be most concerned are those which in this list are largely formal duties. The official trade unions' representational remit, including housing and cultural issues, on the surface appears quite wide, reflecting the fact that the enter-

[33] Khrushchev made an abortive attempt to reorganise the unions along territorial lines as part of his plan to reform the economy on the basis of regional economic councils (*sovnarkhozy*). After the attempt to unseat him in June 1957, however, Khrushchev was disinclined to force this measure on the resistant trade unions and the branch union structure remained.

prise was the basic unit of Soviet society, which provided not only employment, but also housing, medical care and leisure facilities for its workers. Hence the trade union committee was supposed to be concerned with all these questions. As mentioned above, however, these supposedly representative functions were closely bound up with the role of the union in stimulating higher productivity and improving labour discipline. The trade unions' involvement in issues of service and leisure provision was a source of patronage and control rather than an extended frontier in collective representation.

The character of the 'collective agreements' negotiated by the unions requires explanation. There was little room for 'negotiation' of this agreement, since allocation was determined by the centrally determined plan. The union was allowed to comment on the document drawn up by management, and drew up the sections concerned with welfare provision, but since it was only a document of formal significance there was little need for effective consultation.[34] The trade unions were also supposed to monitor the implementation of the collective agreements and health and safety rules and to mediate in disputes. This in practice did not occur, however, because the trade union was the weakest member of the factory triumvirate composed of itself, management and the Party. It was heavily dependent on management for the goods and services it controlled, and in terms of the career structure its officers were part of enterprise management. Therefore, in practice the union officials had no interest in creating conflict within the enterprise. Meanwhile, workers, since they did not see the union as existing as a channel to represent their interests, did not exert any pressure on their unions to defend them collectively, though they might seek the help of an efficient union official over personal problems such as housing, debt and so on.

The unions were supposed to represent the grievances of individual workers. The union was, for example, obliged to give its agreement to any dismissals of workers proposed by management. McAulay estimates that in Leningrad in the period of her study (1957–65), the unions agreed to half of all management requests for dismissal. The reason generally given for refusing assent was that the worker should be given another chance, rather than because there was no legal basis for dismissal (McAulay, 1969: 123). Such defence, however, took place within the parameters set by the Party. Thus the unions did, to a limited extent, defend workers' interests, but only as these were defined from above by the Party and not as expressed by workers from below. In the case of dismissals the trade unions' attitude was governed not by a commitment to defend workers, but by the interests of society as a whole, as

[34] Since conditions of employment were outside the scope of the collective agreement, the Party periodically attempted to make it serve other functions. As has been seen, in the early Stalin period it was hoped that collective agreements could serve as a means of improving productivity, while in the Khrushchev era the enterprise collective agreement was supposed, according to the December 1957 Central Committee resolution, to be one of the ways in which trade unions could participate in management (Ruble, 1981: 34).

represented by the Party. From the Party perspective, reasons for keeping a worker included the labour shortage, and the potential social disruption that could come from leaving undisciplined workers outside the care and control of the labour collective. But, as McAulay notes, if a worker was having a bad influence on others (mainly by persuading them to drink), or worked in a young collective, the trade union was very unlikely to defend him or her: production was the Party's number one priority. The enterprise trade unions were rather toothless watchdogs over the implementation of the central Party policy, because they could not assert themselves without the backing of the enterprise Party committee. Such backing would most often be given in the context of a particular Party campaign organised from above or in the event of conflict within the enterprise administration (Clarke *et al.*, 1993: 112).

The unions had a similar level of effectiveness in their role as promoters of productivity. As indicated above, they were supposed to act to break the collusive relations that developed between workers and managers seeking merely to negotiate an achievable plan and meet it, thus supposedly acting in the interests of the working class as a whole by raising production. But again the unions did not have the requisite power to challenge the established industrial culture of enterprises: they were too weak in relation to management and the Party to pursue this abstract 'interest' of the working class. Rather, they collaborated with the local Party and enterprise administration to ensure that the prevailing form of relations within the enterprise were not challenged. The enterprise unions were therefore at the same time constrained by their formally prescribed role and unable properly to fulfil it.

Soviet trade unions in comparative perspective

The main difference between trade unions in the West and those that existed under communism is that the former were created by workers to defend themselves against their employers whereas Soviet trade unions were subordinated to the state. Their main task was to serve the state by promoting production and ensuring order within the enterprise, and their scope for defending workers was strictly limited by the priorities of production.

Nevertheless, in order to avoid constrasting Soviet trade unions with a hybrid, healthy form of trade unionism assumed to exist in the West, it is important to recall that Western trade unions are not necessarily the picture of institutional health. This is highlighted by the fact that debates about the relationship between unions and their members in the West tend to revolve around the question of how far union organisations develop an institutional interest distinct from that of the members they are supposed to represent. Variations of this debate have been current ever since Michels, writing about the German Social Democratic Party in 1915, asserted the 'iron law' that 'organisation implies the tendency towards oligarchy' (Michels, 1915: 37): as politics became more complex, so the emergence of a caste of professional leaders became a technical and practical necessity. More recent formulations of the debate see the central tension within unions as being between hierarchical accommodative forms of unionism and workplace-based activity (Fairbrother and Waddington, 1990: 16), which expresses the fact that unionism functions both as part of and in opposition to the labour–capital relation (44). That is, as Hyman has put it, 'external constraints – the power of employers and the state – impose forceful limits on the purposes adopted by trade unions. They find themselves accorded legitimacy, recognised and even encouraged, only when their aims and actions do not seriously challenge the continuation of capitalism' (Hyman, 1975: 92). Moreover, the institutionalisation of unions through means such as the closed shop, 'creates mere paper trade unionists. . . . They relate to the union passively, as atomised individuals and *not* as participants in a living collectivity. Not surprisingly, "the union" represents a distant and impersonal power not an expression of their own identity and interests' (Hyman, 1989: 179). The paradox of trade unionism is therefore that 'workers' organisations which are defined and constituted through struggle tend also to contain and inhibit such struggle' (Hyman, 1989: 181). But, according to Hyman, the containment of struggle is never complete and is always balanced by the '*practice* of workers themselves' (Hyman, 1975: 92).[35]

The main point to be made about this debate is that, whether the gulf between unions and their members under capitalism is emphasised or minimised, what is at issue is the tensions which exist in union organisations. Such tensions arise because unions have their own institutional interests and vary in the degree to which they involve their members, but the important point is that without their members they would have no *raison d'être* at all. However bureaucratic a union organisation might be, it is therefore always subject to some kind of pressure from its members. It is this tension which was absent from Soviet trade unions: they were never constituted to represent their members' interests. For example, Sweden has a highly centralised and disciplined form of trade union organisation which has superficial similarities with the Soviet unions. There have been moves towards more decentralised bargaining since the 1980s, initiated by the employers, but prior to this bargaining was carried out centrally between the Swedish trade union confederation (LO) and employers' organisation, SAF. In the 1940s balloting on collective bargaining in Sweden was abolished (Kjellberg, 1992: 90), and unions have generally been expected to show 'social responsibility' in exchange for social reforms and improved material conditions (95). In the late 1960s this led to mounting dissatisfaction among workers, which expressed itself in a wave of wildcat strikes. Such action

[35] Such formulations have not gone unchallenged. It has been pointed out, for example, that far from always being sold out by compromising leaders, union members sometimes prove to be more conservative than their leaders, as in cases where they refuse to take strike action (Kelly, 1988: 156). Others, meanwhile, have criticised the ethos of 'workerism' or 'rank-and-filism' which they claim informs the critique of bureaucratisation, and have noted that 'the extent to which worker interests can be promoted without formal organisation is limited' (Batstone, 1988: 86).

showed the extent to which the relationship between the unions and their members had weakened, but it also demonstrated that the loyalty of union members could not be taken for granted forever: as one commentator has argued, 'the success of Swedish corporatism depended on the strength of the labour movement and it was this very strength which meant that internal discontent became expressed in a strong radical surge' (Fulcher, 1987: 249). Corporatist arrangements in Germany faced similar problems. The trilateral co-operation of the *Konzertiete Aktion* (concerted action), which lasted from 1967 to 1977, gave the unions influence, but the consensual wage policy also limited the extent to which they could pursue their members' immediate interests. *Konzertiete Aktion* therefore came under heavy pressure from rank-and-file union members who could see no benefits in return for their voluntary wage restraint (Jacobi *et al.*, 1992: 239). In contrast to this, Soviet trade unions faced no pressure from their members, who did not *expect* their unions to defend their rights.

Was the role of Soviet trade unions functionally different from that of the German and Swedish unions at a central level? Jerry Hough has shown that Soviet trade unions did pursue certain distinct policy objectives in their discussions with state bodies. For example, in the discussions for the 1977 constitution and the five-year plan of the early 1980s, alongside their ritual statement of the importance of improving productivity, the unions pointed out that the best means to achieve this was through wage increases, improvements in housing conditions, educational and cultural measures, and improvements in services and safety measures (Hough, 1979: 376). It could therefore be argued that workers were 'represented' within the framework of Soviet institutional pluralism as they would be within a neo-corporatist framework. Soviet trade unions were able to provide information and advice to the branch ministries and Goskomtrud (the State Committee for Labour and Social Questions) and to liaise with Gosplan (the State Planning Committee). This is superficially similar to the situation in Sweden, where consensus developed between the state and the major interest organisations, which co-operated to maximise economic growth and international competitiveness, control society and divide up the national product (Crouch, 1979: 187; Lehmbruch, 1982: 17; Schmitter, 1984: 104, cited in Fulcher, 1987: 232). It has, moreover, been argued that the Swedish labour movement was integrated and subordinated through this 'corporatism' (Panitch, 1980: 12). But the crucial distinction to be made is that, while it can be argued that the independence of Swedish unions was undermined by their co-operation with employers and the state, the leaders of the LO, as the grass-roots rebellion of the late 1960s reminded them, were ultimately responsible to a distinct constituency. Their co-operation with employers at a central level depended on their ability to deliver the support of their members for any decisions they made. In contrast, the influence of the Soviet trade unions at a central level was granted by the state, and the unions' interventions were only indirectly influenced by their supposed constituency.

Soviet trade union members were thus 'paper trade unionists' *par excellence*, who did not look to the unions to defend their rights. The 'practice' of Soviet workers therefore did not exert any direct pressure on the unions to change. This disconnection from their membership has continued to plague the trade unions in the transition era and is one of the reasons why the unions have been unable to break out of their dependence on employers and the state.

The case study

This chapter began with the puzzle of workers' passivity in the face of the assault on their living standards in the post-communist period. My case study was designed to illuminate why workers' reaction to reform had been so muted. I examined this question from two perspectives. First, I asked why it had proved so difficult for the former communist trade unions to transform themselves into independent organisations capable of representing workers' interests during the transition period. Second, I examined why, given the limited extent of reform within the unions, workers had not organised within or outside existing structures to defend their interests?

Research strategy and choice of research site

My questions implied detailed examination of social processes, for which ethnographic case study research is ideal. One objection frequently raised against the case study method is the alleged impossibility of knowing whether the processes discovered in one case are more widely dispersed. Taldym is an enterprise with a history of worker and trade union activism in the most militant branch of industry, coal mining. It was selected as the object of research precisely because it is atypical, and thus serves as a 'negative case' to test my underlying hypothesis that the failure of workers' organisation in Russia cannot be explained voluntaristically, by reference to the inadequacies of either workers or trade union officials, but has deeper structural roots. In contrast to the situation at most other enterprises, workers at Taldym have shown themselves capable of mobilisation and the trade union president is committed to change. But like other workers, those at Taldym have proved unable effectively to defend themselves during transition. Since this failure cannot simply be explained by 'passivity', the case of Taldym highlights the structural barriers to the representation of workers' interests which result from the Soviet institutional legacy.

Mineworkers had been one of the most active groups of workers in the *perestroika* period and after, being the first group of workers to launch a nationwide strike in 1989, which they quickly followed up with another major strike in spring 1991.[36] These strikes had begun spontaneously and were led by

[36] For accounts of these strikes see Clarke *et al.*, 1993; Friedgut and Siegelbaum, 1990; Mandel, 1991; Rutland, 1990; and Teague, 1989.

independent strike committees which grew up during the course of the action. Not only had the mineworkers proved themselves to be the most active in Russia, but as a result of their action the former communist miners' union, Rosugleprof, also took up a position as one of the most outspoken of the former communist branch unions. In the wake of the strikes, there was considerable renewal within the union, with members of the strike committees being elected to replace members of the old mine union committees. The miners' union also had to face competition after the formation of the rival Independent Miners' Union (NPG), which claimed genuinely to represent workers' interests. For these reasons Rosugleprof was assumed to be the official union most likely to be showing signs of reform and therefore a suitable focus for a study of the unions in transition.

Having identified mineworkers and their former communist union as the subject of my enquiry, the next step was to find a suitable site for the analysis of their behaviour at enterprise level. In the first stage of research, in the spring and summer of 1994, a variety of South Kuzbass mines around Novokuznetsk, Osinniki and Mezhdurechensk were visited and interviews were conducted with mine trade union presidents, other union officers and workers, as well as with regional union officers. This provided an overview of the working of the unions at a mine level,[37] as well as the basis on which to select a mine for a detailed case study. I had initially intended to analyse a mine organisation of the NPG as well as a Rosugleprof trade union committee, but the interviews with NPG leaders at mine, town and regional level conducted during this phase of research convinced me that the NPG was a corrupt and dying institution, which had lost touch with its few remaining members.[38] I therefore decided to concentrate my analysis on Rosugleprof, to which the overwhelming majority of mineworkers still belonged.

Since I wanted to analyse the barriers to the reform of the former communist unions I needed to choose a strong 'negative case' where the union committee was active and committed to reform. If such a union had not managed to transform its mode of operation, this could not be explained by personal factors such as laziness or inertia on the part of the trade union president, and the case would thus highlight the structural constraints on change within the union. Taldym was chosen because the initial phase of research revealed the union to be one of the most radical and active in the region. In contrast to many of his counterparts at other mines in the region, the mine trade union president is a very energetic man who tries, in his own terms, to do his best for his members. He has a good reputation within Rosugleprof and is locally viewed as a strong and intelligent leader. He has repeatedly been offered posts within the

[37] The conclusions of this stage of the research are presented in Ashwin, 1995.

[38] The corruption and co-option of the NPG is an interesting issue in itself, and one clearly related to the problems of workers' organisation in the transition period. But the problems of the independent trade unions have already been analysed by Clarke *et al.*, 1995, and Ilyin, 1996b. By the time I got to the Kuzbass there was very little left to say.

South Kuzbass territorial committee of the union (which he has turned down). He usually plays the leading role in promoting union action in the South Kuzbass and is highly critical of the conservatism of his counterparts at other mines, claiming that most of them have no leadership qualities and are unfit for the posts to which they have been elected. At a national level the Taldym president is also a prominent figure. He is a member of the national tariff agreement commission, so he is well informed about union policy and makes regular trips to Moscow. As one worker from the mine put it, 'in Moscow he's *na ty*': able to refer to his superiors in the informal 'you' form, a sure sign of respect. In Moscow the Taldym president is known for his radicalism: at the 1996 quinquennial congress of Rosugleprof he publicly criticised the former union president, Vitalii Bud'ko,[39] for his timidity, and is generally critical of the union leadership. The Rosugleprof mine president is proud of his record; as he said during the preparations for an (abortive) strike in August 1996, 'everyone will be looking at Taldym to see what we decide; we always lead the way'.

Not only is the union president at the mine unusually radical, the Taldym workers are some of the most active in the South Kuzbass. The mine was the site of one of the first major strikes in April 1989 and was the first to join the general miners' strike which started in Mezhdurechensk in July 1989, and the Taldym workers refused to settle after the miners in Mezhdurechensk had gone back to work. The proven potential for worker mobilisation at the mine and the activism of the trade union president make Taldym a particularly appropriate and interesting place to investigate the nature and limits of change at the enterprise level.

A small NPG organisation grew out of the strike committee at the mine, but it has never attracted more than 300 members. (Its membership has fluctuated between 200 and 300.) When asked about the NPG in interviews workers often display only a hazy knowledge of its existence, and they are almost always ignorant of the principles which distinguish it from Rosugleprof.[40] The Taldym trade union committee has thus proved robust in the face of competition: the president has tended to regard his NPG counterpart more as an irritating fly than a disturbing thorn in the flesh.

The specificity of the coal industry

Coal mining may appear to be a suitable negative case in terms of worker militancy, but does it differ from other industries in too many other ways to be an appropriate site for a case study? Mining communities have long been held to have specific features which promote solidarity (Bulmer, 1975), but detailed

[39] At the beginning of May 1998 Bud'ko was offered and accepted a government position as Chair of the Federal Committee on Regulation of Natural Monopolies and Transport.

[40] NPG presidents at the mine level generally see the defining feature of their union as the fact that they do not allow managers in their ranks. This does not, however, render them immune to the structural pressures to co-operate with enterprise management.

studies of union formation in the mining industry, such as that carried out by Huw Beynon and Terry Austrin (1994) with regard to the Durham miners, have shown that the solidarity of miners is far from automatic. Moreover, as will be shown in the next chapter, the differences between mining and other communities are less significant in the Russian context because of the uniformity of the social form of the labour collective imposed during the Soviet era. This defined a certain homogeneity of social relations across sectors and regions,[41] and although differentiation is now increasing as the economy changes, the legacy of the past is still tangible.

The more pressing question concerns the fate of the industry in the post-Soviet era. In the communist era, the relationship of labour collectives to external agencies – the branch ministry and state planners and ultimately the Central Committee – was a crucial determinate of their internal politics, since these conflictual relations were what defined the common interest of the labour collective. In the coal industry these relations have continued to exist in a modified form. For the duration of its existence (1993 to 1997) the Russian Coal Company, Rosugol', played a role akin to that of the old branch ministry, and coal mining was dependent on state subsidies, so the link with the state was preserved. (The functions of Rosugol' have now been assumed by the Coal Department of the Ministry of Fuel and Power.) The situation differs somewhat in other branches of industry where branch ministries have been abolished and subsidies are no longer paid. This contrast is discussed in more detail in the conclusion. For the moment, however, this distinction has not had a major impact on the form of activity in other branches. The vast majority of strikes since 1993 have been in coal mining and education: precisely those sectors where the link with the state has been preserved. It may be that new forms of worker activism are gradually emerging in industries which are not subsidised, but so far there has been no significant industrial action during the transition which has not been in those industries where the state is still involved. This underlines the point that breaking out of old patterns of relations is difficult even where some of the key structural constraints which defined those relations are removed.

Methodology

After the initial phase of research in the spring and summer of 1994 I made three month-long research visits to Taldym, in October 1994, June 1995 and August 1996. I maintained regular contact between visits with my key informants by letter and telephone and I returned to Taldym for a week in June 1998. In line with the structure of the book, there were two main strands to

[41] As Ilyin argues when introducing his case study of a town bus company, 'In essence all state enterprises are alike. Therefore it is possible to use a case study . . . to say something about the essence of social relations, contradictions and conflicts in state enterprises throughout Russia' (Ilyin, 1996a: 64).

the research at Taldym: work with the trade union and research among the workers.

Research with the trade union involved repeated interviews with the trade union president and vice-president and members of the presidium; interviews with shop trade union presidents; observation in the trade union office; observation of meetings of the trade union committee; and observation of shift meetings. I attended important annual events such as the meeting to conclude the collective agreement (in 1995), the celebration of the 'Day of Trade Union Workers' (in June 1994) and 'Miners' Day' (in 1996), and was also taken by the trade union to see the facilities which it helps to administer, such as the kindergartens and the prophylactic care facility.

I developed a good relationship with the trade union committee, who were very welcoming and allowed me unrestricted access to their office. Amid the bustle of the office my presence was quickly forgotten: the fact that I witnessed a range of reactions to members on the part of the trade union officers – from solicitous and compassionate to impatient and brusque – indicates that they were responding to particular individuals and situations rather than concentrating on projecting a certain image for my benefit. The trade union president positively encouraged me to attend trade union meetings while I was at the mine, and made no objection to my taping them. In short, I was offered a great deal of assistance, which was not accompanied by any attempt to control my movements: the trade union president was quite happy for me to wander around the mine talking to whomever I wished.[42]

I also conducted a limited number of short interviews with members of the mine administration (as well as with the head of the settlement administration) in order to gain specific pieces of information. Generally, however, managers were far more suspicious than the workers and union officials. They answered questions very generally, or not at all, and would often spend a long time grilling me about the source of my research funding. They were clearly perplexed by my presence, and seemed to find it hard to believe that I had a purely academic interest in their mine. I did not want to encourage anyone to complain to the director about my research, so I limited my contact with senior members of management to a bare minimum. (I did, however, have more contact with line managers.)

My research on 'workers' focused on work collectives rather than individual workers. During the initial phase of research I conducted a number of group interviews with both surface workers and miners at Taldym. Notwithstanding the comparative activism of the mine trade union committee, the workers had been almost universally scornful about the 'representation' it offered, and I was thus concerned to discover whether work groups had developed informal mechanisms to defend their interests in the absence, as they saw

[42] My initial contact with the mine was through the trade union. It is an indication of the influence of the union committee that, when I decided to conduct a more detailed case study at the mine, they simply informed the director of my intentions and he left me to my own devices.

it, of adequate formal channels of representation. This research question led me to focus on immediate work collectives in order to discover how they dealt with day-to-day conflicts; to examine whether the 'collectives' themselves constituted defensive alliances; and to find out whether the shop trade union committees played a greater representational role than workers claimed in interviews. My strategy was to select a number of collectives for detailed research in order to analyse both their internal relations and the way that they related to management and the trade union. I also sought more general information about workers' lives, and this was easier to process in the light of the understanding I developed of their collectives.

The collectives selected were all from auxiliary shops, although a selection of individual miners from both production and development shops were interviewed outside work. This focus was partly determined by an interest in 'women's collectives' in the mine, which have so far received very little attention in the research on Russian mineworkers, and partly by ease of access: it was possible to sit for long periods observing the work of surface workers (and maintenance workers who work both on the surface and underground), something which was precluded in the case of miners.

The 'women's collectives' selected for detailed research were the *lampovaya*, where the miners' arrival at and return from work is recorded and their lamps are recharged and serviced; the *zaryadnoe depo* (locally known only as the *zaryadnoe*), where the batteries used for in-mine transport are recharged; the central *kotel'naya* (boiler house), which provides heat and hot water for Vishnovka; and one of the mine's two 'technical complexes', the Vishnovskii complex, where the coal is improved by the manual removal of lumps of rock as the coal passes on a conveyor. The 'men's collectives' studied were a brigade of fitters from one of the in-mine transport shops, responsible for the maintenance of the conveyors in the mine; a brigade of fitters responsible for repairs in the *kotel'naya*; and one of the brigades responsible for the 'modernisation' and adaptation of newly acquired mine machinery and the upkeep of the old in a shop known as URZO.

These collectives span a range of statuses in the mine. The *lampovaya* is one of the 'best' 'women's collectives' at the mine and competition for jobs there is stiff. The technical complex, meanwhile, is one of the lowest-status workplaces in the mine: the work is highly unpleasant, while the women who work there have a reputation for being 'rough'. The *zaryadnoe* is situated between these two extremes, and the *kotel'naya* is positioned just below it. Comfort, ease and cleanliness are the basis for the gradations between women's collectives:[43] there is very little variation in the pay and skill grade allocated to them. In contrast, the variations in the pay of male manual workers at the mine are far greater. A face-

[43] Daily discomfort is accorded far more significance in these gradations than long-term health risks. The gases emitted by the batteries of the *zaryadnoe* contain cadmium at several times the safe level, but this is still seen as a more desirable workplace than the technical complex, where the work is harder and dirtier, but where the coal dust is less noxious than the gas of the *zaryadnoe*.

worker in one of the most successful shops can earn nearly four times as much as the lower-paid male surface workers. The male fitters from the transport shop have a similar status to their female neighbours in the technical complex, though in a male rather than a female hierarchy. This is an insubordinate collective, which is low paid in comparison with other male collectives. The workers of URZO, meanwhile, work in what Russians refer to as 'a warm place', where the money is good, the labour turnover is low and the work is, according to those who do it, satisfying and creative. There is little tension between line managers and workers in this collective, although this is the collective from which the president of the strike committee formed during the 1989 strike was drawn.

There were two sides to my work with the collectives. One was to visit their workplaces, to sit and observe their work and their interactions with each other and with their managers, and, during breaks, to conduct short, informal interviews or simply to chat with workers.[44] The other was to interview certain members of the brigades (work groups) outside work in greater detail.[45] In a number of brigades I developed close relationships with workers from my chosen collectives, who invited me to their homes.

The material I gathered during these visits was invaluable: respondents were not surprisingly more relaxed and confiding in their own homes. I used a variety of techniques during these visits: I conducted taped interviews (with those workers who would allow it), had long informal conversations, and observed the interactions within the family; I was also shown family photo albums, certificates of merit from work, medals and other such mementoes. During the first interview I would usually obtain brief work and life histories, and ask general questions about life and work in the settlement. Then in subsequent interviews I would follow up on specific issues, arising either from my analysis of earlier interviews or from my knowledge of current affairs at the mine.

Those who were happy to be interviewed outside work and invited me to their homes tended to be the more self-confident, gregarious members of their collectives; they were usually 'good talkers' with strong opinions. Obviously, this had some influence on the nature of the material gathered, but since one of my themes was why there was so little overt resistance, working with the boldest and most outspoken members of the collective made sense: such workers acted as 'negative cases' of my evolving hypotheses about the forms of dependency fostered within the mine. Moreover, such workers also tended to be

[44] My visits to the collectives were generally welcomed by workers, although some collectives were more open than others. My first visit was usually disruptive: all the workers would gather round to talk. On subsequent visits I generally received less attention and it was easier to observe and conduct more specific discussions with a few workers rather than the whole group. I felt very welcome and I was able to observe the ebb and flow of life in the collective: encounters with managers, political arguments, discussions about the previous night's soap opera and so on.

[45] For details of the evolution of the brigade system in Soviet industry see Slider, 1987.

the informal leaders of their work groups, and their normative conceptions of 'the way things should be' tended to be those which dominated in their collectives.

My sex also obviously had a strong influence on the process of developing such relations with workers. While women were able to invite me to their homes, men were not. First, such an invitation would have had a hint of impropriety to it and, second, Russian rules of hospitality decree that guests who are not close friends must be well fed: thus while women can make impromptu invitations, men are not at liberty to do so without knowing whether their wives will be available, and prepared, to cook. For this reason, my interviews with male workers were gathered in a slightly different way to those with female workers. In both cases I interviewed workers within their collectives, but, since there are no suitable public meeting places in Vishnovka, I had less chance to spend time with members of my chosen 'men's collectives' outside work. Therefore, in order to gain a deeper sense of the way in which women workers differed from their male counterparts, in addition to my work at the mine, I conducted longer interviews at home with the husbands, sons and male family friends of my close female respondents.

The structure of the book

Chapters 2 and 3 introduce the case study through an analysis of Taldym and the settlement in which it is situated, Vishnovka. Chapter 2 describes the community and highlights the role of the mine within it, indicating how far the enormous influence of the enterprise over workers' lives is a feature specific to small mining settlements and how far this results from the structure and organisation of Soviet enterprises more generally. In addition to this, the chapter also analyses the way the community has been affected by reform, an appreciation of which is crucial to understanding the prospects for union reform and worker organisation within the mine. Meanwhile, Chapter 3 describes the structure and organisation of the mine, concentrating on the impact of privatisation on the mine. Taken together these chapters contextualise the case study, providing an essential sense of the way in which the mine and community operate.

Chapter 4 is a case study of the attempts of the Taldym union committee to adapt to its changing environment. The analysis of the union at Taldym exemplifies the dilemmas faced by the enterprise committees of the former official unions in the transition period, and highlights the contradictions of their position after the demise of the Communist Party within enterprises. In particular, it considers how far the constraints on union reform are structural, and how far they are the product of ingrained behaviour and perceptions on the part of trade union officers, the mine administration and workers.

Proceeding from the analysis of Chapter 4, which shows that at the mine level the union still operates in partnership with management, Chapter 5 asks whether the shop trade union committees operate within the same constraints as the mine union committee. The chapter considers what impact the mounting tension in the mining industry is having on the shop trade union structures. Are they playing any role in channelling the conflict or is it passing them by?

Having established that the union does not yet effectively represent workers, the following section of the book considers why workers have not organised outside the union in order to defend their interests. Russian workers generally express a strong attachment to collectivist principles and appear to have a highly developed sense of class consciousness. Their failure to launch any effective collective protest against the rigours of reform is thus particularly puzzling. Chapters 6 and 7 seek to explain this apparent contradiction, arguing that the forms of collectivism sponsored within enterprises actually inhibited rather than promoted workers' self-organisation.

Chapter 6 shows the way in which the structure of the Soviet regime of production, which constitutes the labour collective as a 'supplicatory unity', and the internal dynamics of the enterprise to which the system of state paternalism gave rise constituted a barrier to the definition and defence of workers' specific interests. Workers identify strongly with the ideal of the united labour collective, but they are also acutely aware of the deep division which exists within this supposedly solidaristic entity. Through the analysis of this contradictory consciousness, the chapter highlights the way in which the structures in which they were located constrained the action of workers. The first section considers the form of workers' identification with the enterprise fostered by state paternalism, while the second section goes on to examine the main forms of division within the enterprise and the form of oppositional consciousness to which these gave rise. The chapter argues that what above all characterises this consciousness is that it is negatively defined and is not expressed in any form of workers' self-organisation. The last section then attempts to provide an explanation for this limitation on the basis of the analysis developed in the first two sections. It also asks how far the structural barriers to workers' organisation identified by the chapter are being eroded by transition.

Within the Soviet enterprise, work groups were also constituted as 'collectives', and Chapter 7 asks why it has not been possible for workers to realise the collectivist potential apparently immanent in this form of organisation. It takes one particular collective – the all-women's collective of the *lampovaya* – and examines the changing potential for collective organisation at the grass roots through an analysis of the history of the collective from the Gorbachev era onwards.

The first section of this chapter focuses on the nature of workers' relations to their immediate work collectives. It deals with the question of gender differences in this relationship, arguing that the form of relations found in the *lampovaya*, despite certain specific elements, is characteristic of the mine as a whole. The chapter goes on to chart the development of social relations in this collective from the late *perestroika* period onwards. It argues that, while the collapse of the Party and the erosion of discipline at the mine allowed work collectives more autonomy, workers have not built on the gains made during the strike movement. The reasons for this again relate to the structure of the Soviet enterprise. The relationships of dependency which constitute the enterprise as a supplicatory unity are reproduced within the enterprise, so that individual shops and collectives have a common interest in securing the best deal from the enterprise administration. Thus, in the same way as the director is seen as the 'representative' of the enterprise in the 'outside world', line managers represent the interests of collectives within the enterprise. Workers' reliance on this form of representation inhibits self-organisation within their collectives and also renders them very vulnerable if, as occurred in the case of the *lampovaya* , their 'representative' turns against them.

Chapter 8 considers workers' reaction to change from a different angle: instead of asking why workers are not organising, it analyses the nature of their survival strategies during the transition period and the political implications of these. For although workers do not constitute an organised presence on the political stage, their responses to reform, be they individual or collective, nevertheless play a crucial role in the post-communist recomposition of the Russian state and economy.

The argument developed in the chapter is that workers adopt complementary strategies in the face of reform. On the one hand, they are far from indifferent to the loss of security provided by the social guarantees of the past, or to the destruction of the collective institutions of social and welfare provision. But rather than seeking to build a new relationship between individual and collective, in which the workers would take control of their collective institutions, they remain locked into the alienated forms of symbolic collectivism inherited from the past. Meanwhile, as a complement to their search for collective salvation, workers are also very active in pursuing individual survival strategies. Enterprise collectivism is thus coming under pressure from below as workers increasingly look outside the enterprise for their survival – which in turn reduces the possibility of their mounting a collective response to transition and reinforces their dependence on authoritarian leaders. The complementary perspectives of individualism and 'alienated collectivism' adopted by workers are a key force in shaping the labour collective, and Russia, of the future.

2

Vishnovka: portrait of a mining settlement in transition

Taldym mine is situated in the small mining settlement (*poselok*) of Vishnovka and is the heart of the settlement: the life of the community is inextricably bound up with that of the mine. The following chapter about everyday life in Vishnovka will focus on the role of the mine in the community. It highlights the nature of workers' relationship to the enterprise, and distinguishes which elements of this are specific to post-Soviet mining communities, and those which apply more generally to workers at state and former state enterprises. The chapter contextualises the case study, providing the basis on which workers' preoccupations and aspirations can be understood in relation to their environment. It also analyses the way the community has been affected by reform, revealing the nature of the fast-changing terrain on which the trade union and workers are now operating.

Vishnovka

Vishnovka has a population of 11,500, and is situated to the south of the city of Novokuznetsk. It began to grow in 1930 when the Kuznetzk (now Novokuznetsk)–to– Tashtagol railway line was laid through the village. At this stage the main employment in the settlement was provided by small-scale domestic tool production and a wood processing plant which provided materials to the Kuznetsk construction organisation and the growing coal industry of the region. The construction of the settlement's mine began in 1955, after a 1944 survey discovered large coal reserves in the area, and on 1 March 1957 the mine delivered its first tonne of coal.

The settlement now has a working population of 5,000, and three main enterprises. The most important of these is Taldym mine, which employed approximately 3,000 workers until 1996; by June 1998 the number employed at the mine had decreased to 2,600, though it is still the main source of employment in the settlement. The other enterprises in the settlement comprise an open cast mine, which employs approximately 800 workers, and a motor depot with a staff of approximately 700. There is still a wood processing plant on the outskirts of the settlement, which is a subsidiary of a larger plant in Novokuznetsk, but this is on the verge of closure and now has very few workers left.

The village has two schools, somewhat confusingly known as school number 19 and school number 30, both of which take pupils from 7 years old until school leaving age. Each has 1,000 pupils, who are taught in two shifts, one in the morning and one in the afternoon. The other local services comprise two polyclinics, a music school and, most importantly, three kindergartens. The settlement used also to boast a 'house of creativity' (*dom tvorchestvo*), which replaced the house of the Pioneers in the period of 'depoliticisation' which followed the triumph of Yeltsin. In its 'reformed' form, this establishment used to organise artistic activities for local children and exhibit the results, but it was closed in the first half of 1996 because of lack of funds. The mine meanwhile runs a 'palace of culture' (*dvorets kul'tury*), named 'Progress',[1] which comprises a cinema, sports facilities and a hall mainly used for Sunday night discos and wedding receptions. This establishment is still running, but it is plagued with financial problems, which result in its being periodically deprived of electricity and severely limit the scope of its programme. The mine also has a *profilaktorii* (prophylactic care facility) attached to it where mineworkers can rest and receive therapies intended to inhibit the development of industrial diseases. Meanwhile, down river from the settlement the mine has a 'tourist base' (*turbaza*). The other local agencies are the heating authority, the housing administration department, the water and energy authorities and the funeral and repair services. The settlement has a railway station which connects it with Novokuznetsk in one direction and Tashtagol in the other.

By the standards of the Kuzbass, Vishnovka is considered to be a 'good' settlement. It is sited in the valley of the river Kondoma and is surrounded by hills. On its east side, the landscape is severely scarred by the past activity of the open cast mine, which has not recultivated any of the land which it has mined, but on its other sides it is surrounded by woods and fields. At the bottom end of the village is Taldym mine, situated well away from residential areas. The centre of the village is up the hill from the mine and the railway station. This is the most developed section of the village, where the low-rise apartment blocks of the village are situated and the roads are tarmacked. Here, along the main street, are sited the settlement administration building, the two schools, DK Progress, the building of the former *dom tvorchestvo* and a stall large enough to accommodate about ten market traders. A children's café, 'Forest Fairytale', also stood opposite the main village shop until it burned down at the end of September 1994. The rest of the village, stretching downstream towards Novokuznetsk and away from the river bank towards the open cast mine, mainly consists of small painted wooden houses built by the workers themselves, which are

[1] This is referred to locally by its Russian abbreviation 'DK Progress', as it will be here.

known as the 'private sector'. These are surrounded by kitchen gardens and linked by dirt tracks, lanes and makeshift wooden bridges over streams and drainage ditches. Dotted around the settlement are commercial kiosks, selling alcohol, imported chocolate and an odd assortment of other goods such as nail varnish and tights: aside from the degradation and decay of the social infrastructure these are one of the few visible signs of economic change in Vishnovka.

Housing

Housing in the settlement was in the past provided by the enterprises. To obtain a place to live, workers had to register on a waiting list at their enterprise, which was administered by the official trade union. The building of 'private sector' housing was historically also funded by the enterprises in the form of 50 per cent grants and 50 per cent soft loans to workers for materials. There was no private housing market and no other sources of housing. As has already been noted in the introduction, the housing waiting list was one of the most potent means through which enterprise management and trade unions were able to control and motivate workers. The purpose of the following discussion is to give the background to the politics of housing in the settlement, to explain the variations in the type of housing on offer, and to stress the significance of the housing waiting list in the life of the community which has been the focus of the aspirations of generations of workers and has played a decisive role in the fortunes (or misfortunes) of most local families.

Approximately half the population live in what are known as flats 'with modern conveniences' (*blagoustroennye*), which means that they are not shared and have an internal bathroom and toilet. Just over a third live in the 'private sector' of self-built housing. The mine trade union gave the following breakdown of its employees' housing situation:

- *blagoustroennye* flats – 45.7 per cent
- the private sector – 37.1 per cent
- unregistered – 7.7 per cent
- 'dilapidated/tumble-down' (*vetkhoe*) dwellings (including barracks) – 6 per cent
- *obshchezhitie* (hostels in which families are allocated one room in a shared flat) – 2 per cent
- flats without modern conveniences – 1.4 per cent

Of the mine's employees, 22.8 per cent are on the housing waiting list, 13.7 per cent for an improvement in their housing conditions and 9.1 per cent in order to receive a flat. This situation was broadly replicated at the open cast mine, although the trade union there was unable to provide precise figures.

The conditions of those who live in the flats 'with modern conveniences' are variable. The major variation is in size. The most modern flats are the most spacious. They have an 'improved layout' which means that they have three rooms and a reasonable-sized kitchen. This, in Russian terms, is luxury, for it means that couples with children do not have to sleep in the sitting room, which is the more usual arrangement. In the early 1990s the new trend in enterprise housing provision in the Kuzbass was for spacious brick-built 'cottages'. The open cast mine built enough for sixty families and the motor depot enough for thirty. These are located at the lower end of the village. The mine, meanwhile, built two separate streets of twelve cottages each, in the centre of the village. These houses are by far the best in the village. Most families, however, live in older, one- or two-roomed flats with cramped kitchens. Families with children in the one-roomed flats, of course, get no privacy at all. The two-roomed flats, meanwhile, are only marginally better, as it is often necessary to walk through the main room to get to the second bedroom. Nevertheless, obtaining any kind of flat is a major triumph, especially for young families.

The houses of the 'private sector' are usually far more spacious. The main drawback of the typical private sector house, however, is that it has only cold running water or no water at all, an outside toilet and a separate bath house (*banya*). The bath houses are heated by coal, which provides hot water and enough steam to allow nudity in the depths of the Siberian winter. Washing is accomplished on wooden benches. A private sector house is heated by a coal boiler (*pech'*) in winter. The mine or the open cast mine provides each family of its employees who live in the private sector with seven tonnes of coal a year free of charge. This method of heating keeps the houses warm, although, since they are of wooden construction, there is a high risk of fire. Locally it is claimed that at least one of the settlement's houses is lost in this way every winter. The families who live in the 'private sector' are accustomed to such inconveniences and most have no desire to move into flats. Their only complaint is that running a 'private sector' household is a full-time job in itself.[2]

The 10 per cent who live in dilapidated dwellings, hostels and flats without modern conveniences are obviously in the worst position. These would typically be young families (who constitute the majority of the occupants in the hostels) or workers in low-status jobs without the necessary connections to get priority status on the housing waiting list.

The size of the waiting list masks the real level of dissatisfaction with housing provision. Many workers who wanted to improve their accommodation claimed that they had not put their names on the list because it was pointless: the list was too long. These, for example, were the comments made by miners in a group interview about their housing situations:

[2] Whether a family lives in a flat or the private sector depends most of all on family history: building and running a private house demands specific skills and those workers whose families who have always lived in flats are unlikely to have the requisite experience.

MINER 1: I live in a wooden house in the private sector. It's not in good condition. I have waited fourteen years for a flat. Now they've frozen all building, so now I'll never get one. And soon I will retire . . .

MINER 2: I live in a flat. I have a garage. I live in my mother-in-law's flat. I haven't got my own flat. I've got a dacha, and in the summer I work there in the garden.

How many rooms are there in your flat?

MINER 2: Two. Five people live there: my mother-in-law, my wife and I and our two children. It is hard for us. It is a very small flat. I am forty years old and I want to live in my own flat. I didn't even bother to put my name down on the housing waiting list, because I knew it would be hopeless. I'll never get my own flat.

Such stories are continually encountered in Vishnovka. Dissatisfaction with cramped living conditions, problems with resident in-laws and lack of privacy are all endemic. And so too, as will be seen later, is a burning sense of resentment about the way in which housing is distributed by the mine.

The situation has now begun to change, although for the moment the housing waiting list is still in operation. In 1994 the government ended the housing subsidy to coal-mining enterprises. The end of the building subsidy means that the mine is no longer able to offer new housing for workers to rent, while local enterprises also claim that as a result of financial problems they are now unable to give loans and grants to young families wanting to build their own homes in the 'private sector'. All the accommodation that has been built since 1995 has been offered for sale rather than for rent. The cheapest one-roomed flat for sale in June 1995 cost 28 million roubles – at a time when the best salary of a face worker at the mine was 2 million roubles, and the average women manual workers' wage at the mine was 400,000 a month. The ratio between prices and wages was similar in 1996, but whereas in 1995 the mine had not been offering credit, in 1996 it was offering workers the chance to buy properties over five years through a direct deduction from their pay. Although the properties were being offered for sale, the right to buy was only granted to those at the top of the housing waiting list, so for the moment access is still prior to money in the acquisition of housing in Vishnovka.

The end of 'free' housing was a major scandal in the settlement. In the summer of 1996 the trade union committee was besieged by workers overwhelmed by the injustice of change: 'But it's just not right, if I'd have reached the top of the list the year before last I could have got a cottage for nothing and now I have to pay millions for a one-roomed flat. How can that be fair?' To this the trade union president habitually replied, '52 per cent voted for Yeltsin in Vishnovka. Live this life, you voted for it. Now live it.' For some individuals the pill was too bitter – they refused to exercise their 'right to buy' in protest.

Domestic food production

Domestic food production is a crucial feature of the life of the settlement which shapes both workers' relations to the enterprise and their family lives. This was necessitated in the past less by a lack of money than by the shortage of food in the shops. Workers living in a settlement such as Vishnovka were obviously able to produce food on a larger scale than those who lived in big towns, and they needed to do so because supplies were far worse outside the large centres. It is, however, only in the scale of their domestic production that the workers of Vishnovka differ from their urban counterparts: weekend trips to dachas or allotments were also a vital fixture in the lives of the latter.

Houses in Vishnovka's private sector all have substantial kitchen gardens, and the families who live in such houses often have allotments as well. In the private sector extended families often work together running what almost amount to small farms. As will be discussed later, in the past such production was largely intended for families' own consumption, although more recently it has become an important source of additional income. Such families not only grow a wide variety of vegetables, but often keep pigs, chickens and cows (for both their milk and their meat) in pens and sties around their houses. Other than during winter when the ground is covered with snow, the cows are left to wander freely around the settlement during the day. Animal husbandry involves a great deal of work, for not only must the animals be looked after, they must also be fed. Moreover, workers do not buy hay – they make it themselves in their 'summer holidays'. In general workers who live in the private sector estimated that in the past they produced at least 50 per cent of the food they consumed.

Meanwhile, workers who live in flats almost without exception have allotments where they grow vegetables. These are pickled in the autumn and provide a vital source of vitamins in the winter months. The allotments are allocated by the trade union and in this case there is no shortage. In addition to this, the mine rents large fields from local collective farms in order to allow workers to grow potatoes. The fields are divided up into family plots to which the enterprises lay on buses at planting and harvesting time. Potatoes are the mainstay of the Siberian diet, and both the mine and open cast trade union said that all their workers participated in the potato growing. The supply of potatoes to local shops is very unreliable, so in the words of one of the officials of the open cast mine trade union, failing to plant potatoes would be 'a catastrophe' for a family. In the past, this system was a feature of life at enterprises across Russia, although now some enterprises in more prosperous regions such as Samara are beginning to abandon the practice. Nevertheless, it was estimated in 1996 that 90 per cent of all potato production in Russia came from household plots (Working Centre for Russian Economic Reform, 1997: 104–5).

Domestic production of food is considered to be crucial by workers, who universally claim that they could not survive without it. Thus, alongside the mine the household has always been an important unit of production.

Nevertheless, the domestic and enterprise economies have always been closely intertwined. The mine organises potato production; the mine rents out the land on which workers grow their hay; the mine loans them the machinery required to produce the hay (albeit now at vastly increased prices); the mine clears, prepares and divides up the land into allotments, while in the past it used to provide the money for construction in the private sector.

Social life

The centrality of the enterprise

The mine is the centre of Vishnovka's social world – there is no alternative social focus in the settlement. DK Progress provides various forms of entertainment, but this is an institution run and paid for by the mine. Vishnovka has neither a bar nor a restaurant, while the fire at Forest Fairytale deprived the settlement of its small café. Large-scale events, such as funeral wakes, often take place in the mine *stolovaya* (canteen). Although there are bars and restaurants in Novokuznetsk and the nearby settlement of Osinniki, workers very rarely travelled there in the past, while now the prices are prohibitive, and local opinion is unanimous that such places are patronised exclusively by *Mafiosi* and are too dangerous for ordinary workers. Thus, work provides the main form of public sociability within the settlement. This is not a unique feature of the life of small settlements. The Russian sociologist Sergei Alasheev has made exactly the same point about the centrality of the enterprise in workers' lives in the huge industrial city of Samara: there, too, workers complain that 'after work there is nowhere to go' and see work not just as a source of income but also as a meeting place (Alasheev, 1995b: 70–2).

In terms of the themes of this book one of the most important consequences of the social centrality of the enterprise is that the mine is also the focus of workers' discontent with the leisure facilities on offer. Access to the settlement leisure facilities can be an emotive issue, in particular access to the nearby tourist centre (*turbaza*) which is run by the mine administration. Various celebrations, such as the national 'days' which honour different professions, are held at the centre, which is set in attractive surroundings and has a large dance hall. Work collectives are sometimes granted the right to celebrate events such as the retirement of well-respected colleagues there, and occasionally the administration will throw a party for a brigade there as a reward for production achievements. Such rights are greatly coveted, and resentment is easily generated if it is noticed that one collective is granted more access to the *turbaza* than others.[3] Meanwhile, discontent over the mine administration's use of the

[3] A collective's access to such facilities depends heavily on its line manager's effectiveness as an advocate. Certain collectives have managed to establish a tradition of annual celebrations at the *turbaza*, while others never go there together.

facility is always simmering, waiting to erupt at moments of tension: workers constantly complain that managers and trade union officers use the *turbaza* as their own private party venue, where they drink alcohol paid for by the mine and eat meat provided by the mine's agricultural holdings.

Sex segregation of social life

Men and women inhabit separate social worlds in many different cultures, but mining settlements often preserve a strict sexual segregation of social life even when the wider society in which they are located is moving towards a more mixed pattern of social interaction. For example, the classic account of the life of a British pit village in the 1950s, *Coal Is Our Life*, placed a great deal of emphasis on this, arguing that relationships between work-mates were given priority over 'empty and uninspiring' marriages (Dennis *et al.*, [1956] 1969: 183). Men in 'Ashton' spent their time in working men's clubs and pubs drinking and talking about work, while women were confined to the home: not only was it culturally proscribed for them to drink with men, they also, being economically dependent, often did not have the means to pursue independent social activity. Thus, women's main recreation was meeting for chats over tea with relatives and close friends. And this was not only a feature of the 1950s: in the 1980s mining villages were still reported to have 'a very male character. The social and cultural life is geared largely to those who toil beneath the soil. . . . There are few places where women can meet so their social lives revolve around the kitchen' (Taylor, 1986: 84).[4]

Such segregation is founded on rigid definitions of masculinity and femininity: to be a man is to work, while a woman's prescribed role is to service and stand by her man. These gender identities are partly fostered by the nature of miners' work. Beynon and Austrin have shown, for example, how, in the hard and risky environment of the Durham coal mines in the late nineteenth and early twentieth centuries, a masculine identity associated with ideas of hard work, strength and toughness was adopted partly as a survival strategy, but also acted as a vital disciplining agent: 'Miners . . . taught their sons to endure pain and fear "like a man". . . . Private enterprise coal mining could not have survived without the coal miners whose values, principles and attitudes came to form a central part of the system itself' (Beynon and Austrin, 1994: 152–3). A notion of honourable manhood was vital in keeping the scared 14-year-old boys described by Beynon and Austrin down the pit and in ensuring their elders' exertion while on the job. What this analysis also shows, however, is that while mining is conducive to the forging of a particular masculine identity, the

[4] This description comes from an article in an edited collection about women's role in the 1984 miners' strike (Seddon, 1986). The essays in the collection written by 'miners' wives' who played an active role in supporting the strike tend to stress the social restriction of their lives before the strike: as one Welsh miner's wife put it, 'Looking back, I wonder what I did to fill my time' (Davis and James, 1986: 16).

process is not determined, but occurs within a specific social context. Indeed, the neat association between masculinity and mining first required women to be excluded from underground work, something which occurred in Britain in the mid-nineteenth century.

What is specific about the Soviet and post-Soviet social context in this regard is the high labour participation of women.[5] Since 1970 the proportion of working-age women in the Russian labour force has been close to the biological maximum (Shapiro, 1992: 15), and even though unemployment is now growing, so far, contrary to expectations, women do not form the majority of the unemployed (Ashwin and Bowers, 1997). In the Soviet era, mining areas were no exception to the national trend in female employment: women were employed in surface work in mining, while 'women's industries', such as garment manufacture, were specially subsidised to provide jobs for 'miners' wives'.[6] The position of women in Russian mining settlements therefore differs considerably from that of their counterparts in Britain (or, to be more precise, Britain before its coal industry was destroyed). In Britain, miners' wives generally did not work, and, even where they did, it was not in manual work associated with mining such as coal sorting: the vast majority of the 'pit brow lasses' were excluded from the industry in the 1950s after a protracted struggle (John, 1980: 229). How does the fact that Vishnovkan women work at the mine influence the life of the settlement?

The mass participation of Russian women in paid work has had very little impact on the domestic division of labour. According to Bolshevik ideology, women's labour participation was to be facilitated by a transfer of domestic functions from the private to the public sphere rather than through any redefinition of gender roles, but, aside from kindergartens, the requisite network of institutions remained woefully inadequate throughout the Soviet era. It was this which saddled Soviet women with their notorious 'double burden': women workers continued to take primary responsibility for the domestic sphere. Cooking, cleaning and child-care are still considered by most Vishnovkan men to be *babskaya rabota* (women's work), although food production and animal husbandry are shared by men and women, with men carrying out the most 'manly' tasks associated with this, such as slaughtering livestock and building stalls. The division of labour is less rigid than that depicted in

[5] This is a significant way in which Vishnovka diverges from Bulmer's 'ideal type' of the mining community; Bulmer identified women's confinement in the home as one of the defining features of such occupational communities. It largely conforms to the other characteristics he identified: the physical isolation; the economic predominance of mining; the physically exacting, dangerous nature of the work; the social consequences of occupational homogeneity and isolation; and the close-knit nature of the community (Bulmer, 1975: 85–8).

[6] Such enterprises are no longer subsidised and many of them have now closed. The percentage of women among the unemployed in Kemerovo *oblast'* is therefore higher than in many other regions of Russia, although so far the only group of women to be made redundant from the mines are those who work in *sotskul'tbyt* facilities which have been closed down after being transferred to the local authority.

'Ashton' – because men and women work on different shift systems, men are at least obliged to carry out what the women refer to as 'elementary' cooking, for example – but male performance of such tasks has not eroded the idea that they are women's responsibility.

Does women's status as workers mean that they are able to enter the social world of men? Vishnovka does not have working men's clubs or bars from which women are excluded: the nearest equivalent to a club is DK Progress, and this is open to both men and women. But although women work there is a strict sexual division of labour at the mine and this carries over into social life. Work collectives often socialise together, but male collectives tend to do so on a more regular basis than their female equivalents. While brigades of male workers will often gather informally after work to drink, or go fishing together in summer, women workers usually only meet socially with their collectives at specific events such as retirement parties. Instead they tend to spend free time with close friends chatting over tea or wine.

The frequency of the post-shift drinking sessions of men's collectives varies according to the culture of the collective in question, but they are a feature of the life of most brigades. Indeed, informal vodka-drinking gatherings are the preferred social activity of the men in the settlement. In common with the cultures described by the authors quoted above, such drinking is a strictly men-only affair. Groups of vodka-drinking men are generally excluded from the home and, as mentioned above, there are no bars in Vishnovka. So the men congregate in garages, which are located in clusters some distance from the relevant owners' flats. (In cities such as Kemerovo or Novokuznetsk it is often necessary for car owners to go by public transport to their garages.)[7] Here the drinkers set out their *zakuski* of gherkins, smoked pig fat (*salo*) and occasionally sausage on pieces of newspaper ready to follow up the swiftly swallowed neat vodka. And then they drink, denounce the injustice of life that leaves them thus condemned to the garage, and feel better until such time as they have to confront their wives. This is by no means seen to be an ideal arrangement and although the garage is a valued sphere of male autonomy it is recognised to be a rather demeaning one. As a miner commented in a group interview conducted during a drinking session after the night shift at work: 'It's painful that we live like this. It weighs heavily on our hearts. After work we are so stressed, so we go and drink together because it's the only way that we can relax. Often we drink in each other's garages. . . . You need to drink to get rid of all your bitterness.' The claim that men drink 'to get rid of their bitterness' is a common one, and underlines the fact that, although workers are used to the conditions that prevail in Vishnovka, they are in no way satisfied with them.

A woman worker listening to the above tale of woe, herself married to an alcoholic, commented derisively, 'Stress! You lot don't know what stress is.' Her

[7] One of the advantages of a 'private sector' house is that a garage can be built alongside it. But the garage lots used by flat-dwellers are a much better site for male drinking sessions.

response was typical: not surprisingly the drinking culture causes a great deal of tension between men and women. The men say that what they want is understanding and argue that the 'good wife' is one that allows drinking in the house. This, for example, was the comment of a miner during the same drinking session which was (unusually) conducted in the tiny kitchen of a two-roomed flat:

> We haven't got anywhere to go. There aren't any bars here. If we don't get permission from one of our wives to drink at home, then we drink outside. Lena, she's a good, understanding miner's wife. She's allowed us to be here today. Another wife, she'll shout and swear at us. A good wife understands that it is better for us to get drunk in someone's house than on the street. That way we're safe. Nothing is going to happen to us. She knows where we are.

The wives, however, are under strong pressure to 'shout and swear', because that is what they are expected to do: a woman's role is to disapprove of the drinking and do her best to keep her husband under control. Every good mother teaches her daughter that the biggest disaster likely to befall her in life is to marry a drunkard – and there are plenty of examples to testify to the truth of this. So from their mid-teens onwards conscientious girlfriends try to stand between men and their vodka, which only adds to the other cultural pressures which prevent men and women from socialising and drinking together. The result is that vodka becomes a woman's arch-rival and a man's forbidden fruit. This self-perpetuating struggle only intensifies with time: there are no winners. For what the above quotation illustrates is the paradoxical strength and weakness of both sides. The men are not going to stop drinking, but they will also generally not invade a house against the will of an angry woman. They also candidly acknowledge their frailty while drunk and, though they dislike it, accept that women's criticism of them is just.[8] Meanwhile, women may have the power to prevent drinking in the home, but they can only influence its location. So although there are some occasions, such as holidays, celebrations and family gatherings, where men and women will drink together, the culturally accepted pattern is that men will drink with each other and leave women to worry at home. The model Vishnovkan husband will participate in drinking sessions only about once a month when there is something, such as a birthday

[8] It is a generally accepted fact that men have no control when it comes to vodka – men rarely complain that the attitude of women is patronising. For example, at a trade union committee meeting in August 1996, the trade union president reported on a plan of the mine administration to give each miner a bottle of vodka for Miners' Day (25 August) as some kind of compensation for the fact that they had not yet been paid May's wages. One of the female shop trade union presidents vigorously objected to this plan, saying, 'They'll all get drunk the night before, they won't get home and then Miners' Day already won't be a holiday for their families.' This analysis was completely accepted by the (male-dominated) meeting and the administration was requested not to go ahead with the plan. No one at the meeting suggested that the miners might be more restrained than their critic suggested: indeed, when the shop trade union president later reported what she had said at the meeting to her husband (a miner) he nodded in vigorous approval of his wife's analysis.

of a work-mate, to celebrate. Once a week is considered by women to be tiresome but tolerable. Any more frequently and the man concerned acquires the status of an *alkash* (Russian slang for alcoholic).

So does women's status as workers have no impact on the organisation of the settlement's social life? The most important distinction to be made is that while men and women may socialise separately, women are not excluded from public life as they were in 'Ashton'. In the communist era, women were often active members of the Party, served as shop trade union presidents, worked in the civil defence, or served on the social security delegations. Now many of the organisations in which they participated have been dissolved, but they still serve as shop trade union presidents. Such 'social work' gives women a certain standing in the community. The brigadier in the *zaryadnoe*, for example, was awarded a medal for her social work in the Party and shop trade union committee during the Gorbachev era. This was an enormous source of pride to her, and went a long way to compensate for the difficulties and slight social embarrassment of having an alcoholic husband: whatever he got up to, everyone knew that she was a good and respected worker.[9] Another woman who worked in the ventilation and technical safety shop, the bane of whose life was the fact that she lived in a one-roomed flat with her grown-up daughter, her second husband and stepson, similarly gained a sense of worth from her 'social work' and poured a great deal of energy into it – in spite of the fact that the mine had so abjectly failed to meet her housing needs. She reported cheerfully: 'I have always been involved in "social work". I like doing things for people. Old, young, they all come to me for help. I try to help them all. Before I was elected shop trade union president I was a member of the shop trade union committee. I have nearly always been active. We thought it was good that one of us women should participate in the life of the shop.'

Thus, while in Britain the social separation of men and women was underpinned by the fact that women were excluded from the mine – and, until the 1984–5 miners' strike, its politics – in the Russian mining settlement women can and do participate in the political life of the mine. Women often make outspoken contributions to meetings of the labour collective: it certainly cannot be said of Vishnovka, as Dennis *et al.* said of 'Ashton', that, 'one does not encounter women with ideas about general questions, or interested in cultural activities or the running of organisations' ([1956] 1969: 210). This comment about Ashton wives may well have been unfair, but this is not the point: it would be impossible for a sociologist to make such a judgement about the women of Vishnovka because their participation is so visible. Thus, although in some

[9] Both Beynon and Austrin and Dennis *et al.* note that a woman's fate in the mining communities they describe is largely determined by whether or not she has a 'good' or 'family' man for a husband. This, as noted above, obviously has an important influence on a woman's quality of life in Vishnovka too, but to have a 'bad' husband is not the end of the world: two of the strongest, most popular and active women I knew in Vishnovka had alcoholic husbands.

ways Russian mining settlements appear to be typical 'patriarchal' communities, the reality is more complex than this.[10]

Young people

The young people of Vishnovka are probably no less well catered for than the teenagers of the typical British village: within DK Progress they have more facilities than would be provided by a village hall. The Sunday night discos at DK Progress are not a great source of diversion, however, since, as one young miner explained, the DJ is an old woman who 'only plays records that nobody likes'. The lights do not work properly and the record collection is severely out of date. Again, the gender division persists. Both young men and women claimed that only 'bad' girls went to the disco, although there was no stigma attached to boys who went there. Asked what defined these girls as 'bad', teenage girls said it was the fact that they drank and smoked, whilst the young miner quoted above said coyly, 'they are the sort of girls who, if I took them home, would let me do whatever I wanted with them'. But girls can go to the cinema and use the sports facilities without their reputation being tarnished. In any case, the most popular pastimes have nothing to do with DK Progress: swimming or sunbathing by the river in summer; sledding or skiing in winter; chatting on benches or in kitchens; wandering the streets and going on the occasional expedition to the taiga in summer.

'Nowhere to go'

How do workers feel about Vishnovka's social life? The lack of 'somewhere to go' is not a major preoccupation because workers are used to the social constraints of the settlement. Nevertheless, as already indicated, the drinking culture of the village is much lamented by women, and often blamed on the fact that Vishnovka provides no other ways for men to relax. Men too express a sense of their relative deprivation in occasional moments of laconic reflection. These, for example, were the comments of miners talking about how they spent their free time:

> MINER 1: I can survive, but I can't go out. I am a miner, and in fifteen years of marriage I have never once taken my wife out to a restaurant. Do you understand, not once? The way we live, it's not right.
>
> MINER 2: There's nowhere to go and relax.

A complaint that is made with far greater frequency is that there is no time for relaxation because such a large amount of time is spent on food production in spring, summer and autumn. The claims, 'I have no free time; the only free

[10] This is true of Russian gender relations in general: men and women across the whole of society tend to adhere to essentialist notions of sexual difference, but at the same time Russian women have a self-confidence which is far rarer in a country such as Britain.

time I have is when I'm asleep' and 'winter is a holiday for us – we rest and eat the fruits of our labour' were very common.

Having said this, in a settlement such as Vishnovka expectations are not high and workers certainly value some aspects of the mine's social provision. Indeed, what from the outside may seem small changes have in the past made a great deal of difference to the settlement. In the early 1980s the director of Taldym had been an energetic paternalist interested in raising the cultural level of the village. Under his directorship, the 'Vishnovka experience' became a model of good practice in the Kuzbass and received a lot of attention in the press. The achievements of this era are proudly remembered by Taldym's workers, who often bring him up in conversation. The following comment made by a woman worker at the mine was typical:

> The mine once had a director Aleksandr Ivanovich Shundulidi. He was interested in improving the settlement. He was the first in Kemerovo *oblast'* to initiate the 'aesthetic days' when various artists would come to the schools in the settlement. He did lots of building. He built a fountain and asphalt roads in the village. Before he came to the settlement there was only one big shop – now there are three. There was also only one school in the settlement – he opened another one.

Such innovations were greatly appreciated by the mineworkers, for whom they represented a significant improvement in the quality of life. Seemingly minor developments, such as the building of a children's playground, were seen as major advances by local workers, as the following extract from an interview with a different female mineworker illustrates:

> When we had a different director – Shundulidi – he did a lot for the settlement. It was about eight years ago. Then we had a playground with two paddling pools, one for smaller children and one for bigger children and a fountain. There were lots of pretty wooden things to play on – a roundabout, a slide, little houses. Lots of children used to play there. Parents used to go there too. We could sit there on the benches and read a bit while the children played. It was very interesting there. Now it's been ruined. People don't go there any more. People who live nearby sometimes sit on the benches, but it's nothing like the crowds that used to go there. Before we used to gather and go there. The children used to ask to go there. . . . When the playground was built it was some kind of sensation for the settlement.

Thus, the playground provided a valuable meeting place in the village and generated local excitement. Having attractive things such as a fountain in the village seemed to contribute significantly to local pride. What the quotation also illustrates, of course, is that such achievements can be destroyed. Indeed, the sad story of the playground could almost be a metaphor to describe the fate of Vishnovka itself during the reform period: local institutions have been 'ruined'; people no longer go to local meeting places; the excitement generated by past achievements has waned. Vishnovka was far from perfect in the communist period, but life in the settlement did have some positive features. It is precisely these features that local people feel have been destroyed by reform.

'There's no joy any more': the experience of reform in Vishnovka

> We don't live in hope of the future, but in fear of the future. We live from day to day and look at tomorrow with fear. . . . We don't sing any more because there's no joy any more. . . . The worst thing about present day life is that there's no hope. (Worker, *lampovaya*, October 1994)

This was a typical comment of a worker from Taldym asked to comment on how life had changed during the reform period. People say that they feel insecure, fearful and despondent. Of course, in posing questions about change, there is a danger that people will romanticise the past and feel overwhelmed by the problems of the present. The past is over and done with, so previous insecurity is likely to seem less pressing than present fears. And, obviously enough, people were always younger in the past. Even the consistency of complaints is not necessarily an indication of their veracity: the existence of collective nostalgia for a mythical golden age cannot be excluded. So any attempt to disentangle romance and reality in accounts of change must be cautious and depend on corroboration of oral evidence from other sources. The following section looks at the way in which life has changed in the settlement, in order to give a sense of the context in which the research was carried out, as well as a general indication of what 'reform' means for ordinary people.

Leaving aside the above caveats, what do Vishnovka's workers say about the effect of reform on their lives? The worker quoted above claimed that life had decidedly changed for the worst: people were now 'imprisoned in their own homes'; the country had descended into lawlessness; workers were forced to economise and buy only the basics; their savings had been destroyed by inflation and they were afraid of the mine closing. (This was despite the fact that she personally felt happier than ever before. She was now married to her third husband, who was, she claimed, incomparably kinder than her previous partners.) Her account was almost universally endorsed: other portrayals differed only in their emphasis. The worker's immediate boss in the *lampovaya*, for example, assented vigorously to this catalogue of woe, but in her account of the general degeneration she assigned a special explanatory role to the violent American films which were now broadcast on television and available on video. On different occasions she frequently returned to this theme; she couldn't understand why the government was allowing such pro-violence 'propaganda'. Others placed special emphasis on the fact that, with the establishment of the kiosks, it was now possible to buy vodka at any time of the day or night. The sense of disorientation is summed up by the Russian word that has become the favourite exclamation of the transition period, *bespredel*: no limits. This section will attempt to separate out the various anxieties which make up this general sense of gloom and consider how far these are a response to real developments and threats to the community.

At the most general level, the inhabitants of Vishnovka are anxious that their previously close-knit community is breaking down. People are said to have become closed or reserved *(zamknutye)*, and social life more restricted. 'Now everyone is enclosed in their own fortress' was how one woman worker put it. This was in stark contrast to the memory of a rich communal life under communism:

> We all used to gather, someone would play an accordion, someone a balalaika and we would sing and dance, young and old people together. People were closer to each other then. We all gathered. Families and streets used to gather. We celebrated holidays together. Problems didn't exist. Neighbours helped each other. It was a good time.

The same was true at work:

> Eight or nine years ago the collective was different; we were more organised. We socialised together, celebrated holidays and birthdays together. We even went to the circus together. Not now. People have become closed and aggressive. Before, if one of us had a baby we would all go along and see it. We would help out, if necessary. We went along if someone died.

Such opinions were widely expressed. Many workers, for example, said that they had previously made regular trips to the cinema at DK Progress, but now they stayed at home to watch videos instead. Holidays, it was generally felt, 'were not the same any more'. Such comments obviously contain an element of nostalgia and are not necessarily based on an accurate memory of the social world of the past. But such a weakening of communal ties would be consistent with other trends occurring in the village.

First, economic developments provide a possible basis for such a waning of community spirit. Mineworkers in Vishnovka have never been rich, but what characterised their position before was a strong sense of security. The high level of saving necessitated by the shortages of the Brezhnev era meant that workers felt protected in the case of a rainy day. More importantly, once an individual had secured a position at the mine she would not be dismissed, even if she was no longer required; in this case she would simply be transferred to another section. Moreover, although the social safety net provided by the mine was far from adequate – in particular with regard to housing provision – a variety of discretionary payments or loans was available to those in need. For example, as well as giving soft loans and grants to those who wanted to build their own house in the private sector, the mine would also guarantee the loans of its workers who needed to buy furniture from local suppliers. Such grants and loans allowed young couples set up home without being weighed down by financial worries. In short, the mine was the universal guarantor: it was an (extremely generous) mortgage company, a guarantor of its workers' loans and even an insurance company – workers whose wooden houses burned down were entitled, on the basis of a clause in the collective agreement, to compensation adequate to cover the cost of reconstruction.

Now, although wages are not significantly lower than they were in the past, the mine can no longer provide the same kind of social safety net, while savings have been destroyed by inflation. A general sense of crisis has been created by the fact that workers are now habitually paid late. Until 1995 wages at Taldym were paid only one month late on average, which was a good deal less tardy than at other less successful mines, but since 1996 arrears have begun to mount. Delays of five to six months are now commonplace at the mine. In June 1998 workers received their pay for November 1997 and part payment of December's wage – and this was only after the Trans-Siberian railway had been blocked by miners in May. Although workers are able physically to survive when wages are paid late, this requires planning, hard work and cutting back on non-essential items.

Meanwhile, holiday and sick pay are no longer paid automatically. In the summer of 1996, for example, workers wanting to take leave were requested to sign a document saying that they had agreed to do so in the knowledge that they would not receive their holiday pay in advance. In such situations, the mine trade union and the administration are unable to give concrete information: their answer to the enquiries of disgruntled workers is that they will be paid 'when Moscow sends the money'. That the payment of wages is revealed so starkly to be contingent on the whims of 'Moscow' obviously detracts from workers' sense of social protection. Insecurity is also exacerbated by fears of staff reductions. So far these have been achieved through natural wastage, a prohibition on pensioners working underground, and an anti-drunkenness drive, but few at the mine are confident that reductions will stop at this. In this climate it is not surprising that the social benefits provided in the past have been radically cut back: soft loans and grants are no longer easily attainable from the mine; employment at the mine no longer guarantees credit elsewhere; and the 'insurance' offered by the collective agreement has been rendered meaningless by inflation.

The cumulative result of all these developments is to make people feel dramatically less secure. It does seem that this has had some impact on the social life of the village. In the past, minimal economic differentiation and a strong sense of economic security formed the basis of a culture of open-handedness in which people shared what they had without anxiety as to what tomorrow would bring. This culture still exists, but it appears to be being eroded at the edges. A worker from the open cast mine, for instance, in an attempt to substantiate his claim that 'life was more cheerful [in the past], people smiled more', gave the following example. Previously, if you asked someone for a cigarette in the street, they would give you three or four. The next day you would do the same for them. Now, it had become 'somehow shameful' to ask for a cigarette in the street. He did not provide any explanation of why he felt that was the case.

But it seems likely that this results from the fact that financial worries have become that bit more serious. In the past, if people asked for a cigarette, it was assumed that it was merely because they had not had the opportunity to buy any, and at some point they would return the favour. Now it is 'somehow shameful' to ask for a cigarette, because the request might actually be an expression of genuine need. So whereas giving in the past could be relied on to be more or less reciprocal, now the prospect is arising of some people being perennially, or at least periodically, more needy and less able to give: something which has the potential to disrupt old support networks, rather than strengthen them. This provides some explanation for the widespread perception that people have become *zamknutye* (closed). At least part of the reaction to present insecurity has been for families to look inwards, grow more vegetables, and generally ensure that they are as protected as possible.[11]

Another feature of the transition period which would account for a decline in community spirit is fear. Vishnovka is a small settlement where people tend to know each other, and yet they are terrified of going out on the streets after dark. This fear seems to be a major factor in explaining people's retreat into their own fortresses. This, for example, is the comment of a women worker from the *zaryadnoe* on the effect of fear on her social life:

> I used to go to the cinema. I used to hurry off to see films. Now I have calmed down. I don't go now.
>
> *Why not?*
>
> On Friday someone was killed. He was kicked to death at eight in the evening. It is very dangerous to go out, especially in autumn. There's no snow yet, so it's the darkest time of year. Previously the streets of the private sector had street lights. Then they took them down. Electricity is expensive and none of the local enterprises want to pay for it.
>
> *When did they stop lighting the streets?*
>
> I'm not sure. In about 1985 when Gorbachev came to power, I think. Now we have to light our own house fronts. We have to light the door number until eleven at night and if we don't we have to pay a fine. This gives off some kind of light onto the streets.
>
> The young have become aggressive. It is a result of the fact that there is no programme to give them training and then a job any more. It means that they have to ask their parents even for 500 roubles for a packet of cigarettes. They need money to go out with their friends. And so they begin to steal and so on.

[11] Some mineworkers reported that wage delays had also led to increasing strain within families. Whereas in the past those who lent money to relatives were relatively unconcerned about how long it would take for them to be paid back, this is becoming more of an issue in a period of inflation. Wage delays mean that workers are increasingly required to borrow money, in particular from parents in receipt of pensions, but lending now costs something if the money is not paid back relatively quickly. The fact that generosity is a cultural imperative in Russia only increases the tension surrounding these issues.

Another worker, from the all-female collective of the *lampovaya*, described the same feeling although she attributed it to a different cause:

> Before, I was not scared. We would go out at any time of night. Now at seven in the evening when work is finished I feel nervous coming home if I have been paid . . . I am scared. Perhaps we have made ourselves scared. But I won't go out late now, I'm better sitting at home, unless I'm with Volodya [her husband] . . .
>
> *Why have people suddenly become so scared?*
>
> I don't know. Before the police [*militsiya*] were effective. There was more order. Now there is a Mafia, which this settlement has never known before. They have automatic guns – this was unheard of before. The children's café, near the school, burned down, or was set on fire. Have you seen it? It was about two weeks ago. It was nice, you could eat ice cream there. It was very pretty – wooden and decorated inside. . . . We used to go there with the children. Adults used to go there too, because we didn't have anywhere else. Our settlement is small. No one knows why and how it burned down. Perhaps it was to do with the *kommersanty* [traders] who supplied the chocolate and soft drinks there. I don't know. . . . We don't know if they will rebuild it. But it's in the centre of the village and it will look bad if there's an empty space there.

It may well be the case that a comparatively small increase in the incidence of crime has had a disproportionate impact on the Vishnovkan psyche. The new fear could be in part an expression of shock at the emergence of crime as a problem in a society which, in the pre-Gorbachev era, was remarkably safe. But it is also a response to real changes. The lanes of the private sector, for example, have indeed been deprived of street lights, although not because of the price of electricity. The lights themselves have gradually fallen into disrepair and the settlement council says that it has no budget to repair them – previously responsibility for capital repair was shared between the local enterprises.

Similarly, the position of young people really is changing. Officially, the mine is not taking on any new staff, and so the prospects of young people have become far less certain. In practice, the training of new staff has not ceased completely, and anxious parents who are long-standing and loyal employees of the mine sometimes manage to persuade management to squeeze their offspring in somewhere.[12] Nevertheless, there are ominous signs that a 'lost generation' is in the making. First, the neo-liberals who still hold sway over economic policy are candid about their plans for youth recruitment into the labour market. The 1994 World Bank report on *Restructuring the Coal Industry* simply suggested that all recruitment of new workers to the coal industry should cease

[12] The worker who complained about the lack of a youth training programme initially had problems finding a position for her son at the mine. But eventually the problem was resolved by a visit to the director: 'He [the director] said that he would take him, but as an unpaid worker – a *dubler*. In the end he changed his mind and agreed to pay him. But in general he doesn't want to take on youngsters when he can take on experienced workers from Kapital'naya or Shustalepskaya' (local, less successful mines).

immediately. It was silent on the question of youth training and the fate of the generations thus condemned to a life without work (although the broad prescription of the report is migration away from mining regions – in a country with a chronic housing shortage and no housing market). In nearby Osinniki, which is larger and where the mines are less successful, the results of this policy are already visible. Young people roam the streets, looking like Bolshevik commissars of the 1920s with their peaked caps, leather jackets and tough demeanour. The locals refer to them as 'the Mafia'. Although, there is no such threatening presence on the streets of Vishnovka, the older generations' attitude to young people is changing. Whereas before 'we would sing and dance, young and old people together', now young people are seen as an alien, hostile group; something to be feared. This, for example, was a comment made in a group interview of development workers (*prokhodchiki*), who, along with the face workers, are regarded as having the toughest and most prestigious profession in the area:

> Sometimes the buses just don't come: imagine a miner who lives in Kaltan or Osinniki; he finishes the evening shift and there is no transport to get him home. Sometimes miners have to walk home. After the second shift, the miners who live in Kaltan don't get home until two in the morning. Those who live in Osinniki don't get home until three. It's scary with these bandits and young people roaming the streets. They can attack a miner on his way home.

Of course, the usual distinctions apply: the problem is not seen to lie with the older generations' own family or any young people whom they actually know, but with the 'alien' youths who inhabit the streets. Nevertheless, it is the creation of precisely this sense of separation that is causing the fear. A certain portion of young people are no longer being integrated into Vishnovka's moral universe of the labour collective and so they become outsiders and strangers: not 'that lad who works in the transport shop' but one of those 'bandits and young people roaming the streets'.

As for the 'Mafia' mentioned by the *lampovaya* worker quoted above, the emergence of criminal gangs, protection rackets and various kinds of unregulated trading has been a major feature of the transition period.[13] The public distaste for market traders and kiosk owners means that the label 'Mafia' is applied rather indiscriminately, but that protection rackets are rife is not open to serious challenge. Throughout Russia enterprise directors who fall foul of the local gangs are at risk: the director of one of the South Kuzbass coal enrichment plants was murdered in 1994, for example. The *militsiya*, meanwhile, is generally believed to be incapable of tackling the current crime wave. In the communist era the *militsiya* were very much the poor cousins of the KGB and army. The

[13] It is estimated that three quarters of commercial banks are paying substantial sums in protection money to organised crime, equivalent to half their profits. The murder rate in Moscow is ten times higher than that in the most violent Western European city and as high a murder rate as in New York (*Guardian*, 29 April 1994, quoted in Rose, 1994: 19).

latter two forces kept society under control, and the *militsiya* was never required to deal with major problems of law enforcement. When interviewed in 1994, the former head of Vishnovka settlement administration was candid about the problems. His comments gave some indication of why the local population have such a lack of confidence in their local police:

> There is a 1 per cent local tax to pay for the *militsiya*. But again, the money is sent to Osinniki and we never see it again. In 1992 we managed to buy them a car – before that they didn't have a car. They don't even have a garage. We have no money. The telephones don't even work properly. The *militsiya* is not a prestigious profession. That's part of the problem.

The only real capability of the police seemed to be – as elsewhere in Russia – in the sphere of road traffic offences, where the imposition of fines provides a useful supplement to police officers' income. Given such problems, it is hardly surprising that feelings of insecurity have increased dramatically after the breakdown of communist control structures. And this, as can be seen in the quotations above, has a major influence on the social life of the village. People stop going out and scarce local meeting places are destroyed.[14]

Local government

In addition to this, local services which are part of the *sotskul'tbyt* complexes of the local enterprises face an uncertain future. According to the 1994 Privatisation Law, *sotskul'tbyt* is supposed to be handed over to the local authorities, and many mines have now abandoned their social facilities. Taldym has now given up all of its kindergartens, resulting in the closure of one of them and an uncertain future for the others. The threat to the kindergartens constitutes a real threat to the quality of life in Vishnovka. Quite apart from the fact that they facilitate the high labour participation of women, since children go to school at the age of 7, the first years of schooling also take place at the kindergartens. Moreover, a great deal of effort has gone into building up these institutions – from the tender care of the enthusiastic keeper of the 'botanical garden' at one of the threatened kindergartens to the pride of the cooks in the high quality of the canteen food at the same establishment[15] – and the possible destruction of all this is a very demoralising prospect. Meanwhile, the mine's 'pioneer camp' – a vital resource which used to liberate parents from their

[14] The sight of burnt-out cafés and restaurants is becoming a familiar one in the 'New Russia'. Business people who refuse to acquire 'a roof' (protection) are particularly at risk, although 'protected' premises can also become targets in conflicts over spheres of influence.

[15] The quality of canteen food in Russia is generally abysmal. This is partly due to the poor-quality supplies, but it also seems that workers in this extremely low-status profession are able to give graphic expression to their alienation in the form of their culinary creations. The high quality of the food at the mine's best kindergarten is therefore a tribute to the sense of common endeavour and pride that prevails in this particular labour collective.

children for part of the growing season – has also been closed because the other mines which shared the expense of it were no longer able to bear their costs.

Another disruptive development in the sphere of local service provision is the privatisation of the utilities. Each utility has now been constituted as a shareholders' society (*aktsionernoe obshestvo*, hereafter AO), which means that local government has lost its control over the provision of local services. Prices have increased and the provision has become patchy. Whereas under communism, for example, flat tenants could rely on a constant supply of hot water and heating in winter, now hot water is rationed and in Vishnovka is simply switched off in certain areas for a few hours or days. This is one of the results of the huge chain of inter-enterprise debt that plagues the Russian economy. Under financial pressure the utilities will periodically simply cease supply. While in the past the management of such services would be accountable to the Party, now they are unaccountable. As Vishnovka's former head of administration complained in 1994:

> We are used to communism. Previously people could go and complain to the *gorkom* [the town Party committee] if something wasn't done. Now we, as the municipal authority, don't have anything like the same power. All the services – water, energy and so on – they are all AOs. We don't have any control over them. For example, people complain that the heating authorities don't provide them with hot water but demand payment anyway. I can't do anything about this. I have no rights *vis-à-vis* an AO. Water, the heating service, the repair service – they are all AOs. It is the same all over Russia. Trying to sort things out with the directors of the communal services is very difficult. There is some talk of these services being returned to the control of the municipality.
>
> People come here to me to swear when things go wrong, but there's nothing I can do about it. As they say, 'God is a long way up'.

Similarly, as mentioned above, in the past a variety of local services, such as street lighting, road building and maintenance, was delegated to local enterprises. Now the enterprises, which themselves are under severe financial pressure, claim that they are no longer able to maintain the local fabric. Instead, they pay a local tax, but the head of Vishnovka's local administration claims that the portion of this that they receive is not enough to cover the cost of local services previously provided by the enterprises:

> When the Communist Party existed we divided up things between enterprises – you look after the street lighting, you mend the roads and so on. If the director of an enterprise didn't comply he was sacked from the Communist Party. Now we get our money from taxes. The enterprises pay taxes but we don't receive them directly. We get our money in the form of a grant from Osinniki.[16]

[16] This head of administration had plans for Vishnovka to declare its administrative independence from Osinniki, and to unite with the more prosperous local town of Kaltan. But the plan for independence was defeated after the head of administration in Osinniki, a client of the (then) head of *oblast'* administration, Kislyuk, secured the appointment of a more malleable administrator in Vishnovka in 1995.

> In terms of production we don't look bad on a national level. We make money. But in our local budget we have hardly any money. Now we have problems. The enterprises pay their taxes and we still have to ask them for help. Sometimes they help, but not always. They are all AOs now. . .
>
> In the past there was the stupid *subbotnik* [unpaid working Saturday]. Yes, I agree it was stupid, but it did mean that someone cleared up the dirt once in a while. Now no one does it. I'm not a supporter of the CPSU [Communist Party], but things aren't good now.

The demise of the Communist Party has removed a central component from the old system, and a new administrative coherence and *modus vivendi* are a long time coming. Meanwhile, the economic climate is hardly favourable. With a squeeze on public finance and practically all Russia's enterprises facing major problems as the government attempts to goad them towards real economic independence, Vishnovka's local administration will no doubt face major problems for some time to come.

Political reforms

What of new-found freedom and democracy? Most citizens of Vishnovka have not noticed a significant amount of improvement. Indeed, if anything the political system which confronts Vishnovka's citizens has become, in their view, even more opaque and unaccountable. A constant lament is that in the past people 'knew where to complain', whereas now there is no one to whom they can appeal. In the past, the Party to some extent acted as a trouble shooter and intervened to prevent local tensions exploding into any kind of political conflict. For example, the Party and KGB intervened quickly in the case of strikes or other disturbances, often granting concessions and sacking the enterprise managers or local bureaucrats deemed to be responsible for the problem (Connor, 1991: 222).[17] One worker from one of the transport shops at Taldym mine gave an example of the Party playing just such a positive role in his life. When he was young he lived in a communal hostel with his wife, accommodation provided by the motor depot where he worked as a driver. Six families shared the same kitchen and bathroom, among whom, he claimed, were 'loose

[17] An example of this occurred at a mine construction enterprise in Osinniki in February 1958. A strike was staged after the failure of the bus to arrive at the end of a shift obliged workers, in temperatures of minus 40 degrees, to walk two kilometres back to the pithead. They arrived to find there was neither water nor heating in the changing rooms. Some time after the strike, the First Deputy Minister for Coal of the USSR responsible for mine construction arrived together with the secretaries of the Regional and City Party Committee, the head of the regional mine construction organisation, and KGB officers. The workers had to be coaxed to give their account of events, for fear of repression, and finally one older worker took responsibility for speaking. After what he heard, the Minister instructed the director to go to work as an ordinary miner, so that he might learn how it felt. When he turned up to work, however, he was driven out by the other miners. He gradually clawed his way back into management elsewhere (Clarke *et al.*, 1995: 72).

women' and other undesirable types. One of his neighbours wrote a letter to *Pravda* complaining about workers' living conditions. The letter was published and an order came down from the Party to investigate. Within a week, someone from the Party came round to see the conditions he was living in and was deeply shocked, in particular by the state of the bathroom. The motor depot was compelled to do something about it and he was quickly given a one-roomed flat. This worker's conclusion was precisely that in the past it was possible to complain to the Party and attain some improvement, whereas now such a possibility no longer existed.[18] Paradoxically, therefore, Vishnovka's citizens feel more powerless than they ever did.

This is not to say that the Communist Party was a benign force which protected workers against incompetent bureaucrats. But, unless they became involved in a strike or some other kind of protest, workers were unlikely to fall foul of the communist authorities. And because of its rhetorical commitment to the working class, the Party did find it necessary to intervene occasionally to remedy gross abuses. Certainly, it did not do so with the purest of motives. The occasional very public intervention to sort out a particular problem was good for its image: the workers of Vishnovka still remember such interventions. Timely action by the Party was also an important means of preventing rumbling discontent from turning into anything more menacing.

Now such mechanisms do not exist. The Party has gone, but local democracy is very weak. Local elections in the Kuzbass were not held until spring 1998 because of opposition from the unelected former head of administration in Kemerovo *oblast'*, Mikhail Kislyuk. Meanwhile, Yeltsin's fear of having one of his opponents elected in place of Kislyuk – the popular and charismatic Aman Tuleev – meant that Yeltsin delayed elections for the post of head of administration in the *oblast'* for as long as possible. So, until very recently, citizens in the Kuzbass had no means of influencing local politics other than through organised protest, and even when this occurred it was often not reported by the local media. For example, in August 1996 the area was awash with a rumour that

[18] Such stories are very common. This, for example, was the comment of a worker from Erunakovskii Open Cast Mine who lives in Novokuznetsk: 'Before people could have an influence on decisions at a local level. For example, my mum lives in a village. Under Brezhnev, they got up a petition for a bus to run between the village and Novokuznetsk. They got their bus. Now this would be absolutely impossible. . . . There's *no* accountability now. None. Before it was possible to complain. You could write letters and officials were afraid of getting too many complaints. Since Gorbachev came to power it's just got worse and worse.' A woman worker from Taldym gave a similar example: 'There was a case once when I was on a flight from Moscow to Novokuznetsk. Well, the plane landed at Novosibirsk and they said, "you can't go any further – we haven't got any more fuel – you'll have to make your own arrangements from here". Well, all the passengers got together. It was at the weekend. We wrote a telegram to the Central Committee of the Party and took it to the manager of the airport. Of course, he didn't send it to the Central Committee, but he sent it to the *gorkom*, and they found us a plane double quick and got us to Novokuznetsk. Because they knew that if that telegram got sent they'd be punished and punished hard. The Party didn't forgive things like that. Now there's no one to whom to complain. No one cares about anything any more.'

the head of administration in nearby Osinniki had been arrested at Novosibirsk airport with a suitcase full of money. Pensioners, who had not received their pensions for three months, held a demonstration outside his office, threatening to string him up for making off with their money. The town television station mentioned neither the rumour nor the demonstration, and eventually the head of administration reappeared, without making any comment on the allegations. Whether they had any foundation or not, the incident illustrates the striking lack of *glasnost'* which characterises local politics.

Although the arrival of some kind of democracy is more visible on the national stage, it is not universally viewed as an advance. Workers are sickened by what they see as the venality of government officials and Duma deputies and are deeply disillusioned with President Yeltsin, who is often alleged to be 'just a big Mafia boss'. Although Yeltsin gained a majority in the second round of voting in the 1996 presidential election in Vishnovka, there are very few workers ready to say a good word about him: Lebed appears to be the most popular politician locally.[19] Disillusionment with politicians and all brands of *verkhushki* (bosses) is strong, especially since political corruption is a good deal more visible in the post-communist era. The general opinion is that once politicians get their snouts into the *kormushka* (feeding trough) they forget all about the problems of ordinary people. Disillusionment is also born out of a strong sense that the government is not doing 'anything concrete' about the problems which are afflicting the country. Any satisfaction that workers derive from the fact that they are now able to vote for their political leaders is thus seriously compromised by the disappointment in the results of the policies implemented by the first democratically elected president and parliament, as well as disgust at the conduct of the *verkhushki*.

Despite such disillusionment, the communist political system is not completely romanticised. Many workers do recognise that there was political repression in the past: indeed a high proportion of them are drawn from families who arrived in Siberia after being 'dekulakised'. But repression did not particularly affect workers who lived in the post-Stalin generation. The following extract from an interview, with a female worker in her fifties from the *lampovaya* at Taldym, is a typical response to questions about past repression:

> *Was there any kind of political repression in the past?*
>
> Perhaps the older generation felt it. But I didn't feel anything.
>
> *Could you say what you wanted?*
>
> No. Now we can. We can speak more easily. But among the *narod* [people] repression didn't really exist.
>
> *Well, what sort of things couldn't you say?*

[19] The trade union president and the mine director claimed that the election results which gave Yeltsin a majority in Vishnovka had been falsified, although they had no evidence of this other than the fact that they did not know anyone who had voted for Yeltsin. I also encountered very few people who were prepared to admit that they had voted for Yeltsin.

> I don't know what I couldn't say. What could we say? We live in the back of beyond.

One of her colleagues expressed precisely the same sentiment when she said, 'people weren't scared. They bought their bread and sausage, went to the cinema, and that was all.' Thus, workers had a vague consciousness of limits, but also a strong sense that they – ordinary workers from the back of beyond – were not particular objects of the regime's gaze. Although this speaker could not pin down the nature of past restrictions, however, one clear prohibition of the past was on strikes. As mentioned in the introduction, as late as 1989 the miners who launched the first major industry-wide strike in the USSR's history were aware of what had happened at Novocherkassk[20] in 1962, and were afraid that force might be used against them. Nevertheless, given that ordinary workers in most circumstances were not particularly conscious of political restrictions, the lifting of these is not seen as much of a compensation for the problems of the reform era. The following comment of a thoughtful fitter, again from the *lampovaya*, whose grandparents had ended up in Siberia after being dekulakised, neatly captures this feeling:

> For people who work in TV, who write articles and that sort of thing life, has of course got better. They can write what they like now. They are freer. But for me, for simple workers, life has got worse. Politics never touched us . . . I know that people were in prison, I know it wasn't right. But I knew what to cook then. Now we can't buy any extras.

So although among more thoughtful workers there is an understanding that the end of communism is bringing an end to a particular form of injustice, this does not cancel out the fact that the passing of state socialism has also meant for them a loss of certainty which is hard to bear. Indeed, the prevailing feeling in Vishnovka is that the lifting of the controls of the communist era has simply unleashed a Hobbesian nightmare, a degeneration into a society of robber barons in which only those who can afford protection are safe. A forewoman from the *kotel'naya* gave eloquent expression to this feeling:

> *Perestroika* was supposed to allow us to speak the truth to people's faces – try that now and they'll send you further away than the communists ever did, and that's if they don't just shoot you on the spot. Now we live without laws. You can't go to the *militsiya*: unless you're rich, you won't get anywhere. A friend of mine got burgled and went to them and they said, 'So, what do you want? Look for the culprits yourself.'

Thus, political change has done little to compensate for the economic hardship imposed by transition. This is not least because the 'democrats' have removed accountability to the Party, which constituted an albeit imperfect check on those in positions in power in the past, but have signally failed to

[20] For accounts of this strike, which ended with the army firing on protesters and killing at least seventy or eighty of them, see Haynes and Semyonova, 1979: 73–81.

institute either a 'law-governed state' or democratic accountability at a local level.

A community under threat

The pervasive sense of insecurity and loss of community in Vishnovka therefore cannot be reduced to collective nostalgia for a mythical golden age. It has a real basis. It is partly attributable to the fact that the communist structures have broken down and new and workable structures have yet to emerge in their place. In the case of local service provision, for example, utilities have been privatised, but are not subject to proper regulation, whilst *sotskul'tbyt* is under threat. Enterprises are no longer obliged to carry out capital repair programmes on local infrastructure, but no one else has the money to do it. The home building programme has been frozen. The KGB no longer breathes down the necks of erring citizens, but the *militsiya* is unable to stem the tide of lawlessness. 'Democracy' has been a huge disappointment.

Above all, the uncertainty of the workers of Vishnovka reflects the uncertain position of the mine, which is the centre of the community. The mine no longer pays wages on time; it can no longer guarantee workers' loans; it is gradually ceasing to take on new workers; the threat of redundancy hangs in the air, while the future plans for the coal industry remain shrouded in secrecy. The mining industry is facing a long-term, demoralising crisis and so, by implication, is Vishnovka. Although the precise causes of the crisis may not be fully understood, workers in Vishnovka are increasingly aware that their community is under threat. This, for example, was the comment of one of the miners in the group interview quoted above:

> The government wants to shut all the mines. Can you imagine, in a place like this. Where will the people go? How will we live? There is no other work here. I don't know, there is some kind of struggle for power going on at the top.
>
> In the Far East they are all freezing. They can't import coal from Siberia because of the mad railway tariffs. They have begun to buy coal from China. You think that this doesn't get to us? It gets to us all right.[21] The government isn't doing anything concrete. No one needs us up here.

These fears are by no means misplaced: although the World Bank retreated from its ambitious plan of reducing employment in the Kuzbass coal industry by 70–5 per cent before 1997 (World Bank, 1994: 44), the Russian government, in its efforts to balance the budget, is itself determined to cut the mining subsidy. Amid the financial chaos created by the August 1998 collapse of the rouble it is difficult to predict what will happen to economic policy in the future,

[21] The Russian used here was *'Chto, nam ne obidno? Obidno'*; a powerful rhetorical question and answer conveying an intense feeling of injustice and pain, which is difficult to convey in English. The variant used above is not a literal translation.

but without a radical change in strategy it is likely that the decline of the industry will continue.

The threat to their industry is galling for mineworkers. There is an inevitable sense of rejection and frustration that coal production, which was previously deemed so crucial for the national interest that health and safety standards, social amenities and workers' living standards were all sacrificed to it, is now revealed to be somehow dispensable. What this induces, above all is nihilism. The workers of Vishnovka are simply tired of what they see as meaningless sacrifices and are sickened by the idea of yet more unpleasant, supposedly beneficial medicine. The story of one fitter at Taldym neatly sums up this mood. His father had been a simple miner, an *alkash*. He had drunk 'to get rid of his bitterness', but his generation had also made sacrifices in the hope that their children would live to see a happy future. They had not. 'So', the fitter concluded, 'when Mikhail Sergeevich [Gorbachev] said that we should all work hard so that our children would have a happy future, it was all I could do to stop myself kicking the television in.'

But, however despairing they may feel at times, the workers of Vishnovka have a strong survival instinct: as they are fond of saying, 'hope is the last thing to die'. So, in spite of themselves, they end up, as previous generations have done, working extremely hard in the attempt to secure some kind of future for themselves and their children. The following chapters will look in detail at how they are responding to transition both collectively and individually and consider the impact of their activity on the politics of the trade union, the enterprise, and, ultimately, the nation as a whole. These chapters will also consider how the trade union has been affected by the changing environment described above, and how it is responding to its new situation. The last of these questions will be dealt with first, after a description of the mine and its history during the reform era.

3

Taldym: privatisation and its discontents

This chapter examines the structure and organisation of the mine, focusing on the institutional changes resulting from privatisation. Taldym was one of the few mines that genuinely privatised 'to the labour collective'. At mines that privatised later the form of privatisation meant that a majority of shares were in the hands of Rosugol' and the regional coal concerns, giving these bodies effective control over the strategies pursued in the mines. (Since the disbanding of Rosugol' in 1997 its shares in mines have been transferred to the State Property Fund.) Taldym opted for a closed form of privatisation in which only employees and pensioners at the mine were entitled to hold shares. Both the regional coal concern and managers at the mine have long been attempting to acquire a controlling stake at the mine, but as of 1998 neither group had managed to do so. This means that, unlike their counterparts at other mines, workers at Taldym have had a genuine opportunity to influence the direction of the mine; they cannot be said to have been constrained by the institutional terms of privatisation. But workers at Taldym have only been able to realise this potential control to a very limited extent. This reveals that the terms on which privatisation was conducted are not a sufficient explanation of the failure of workers across Russia to increase control over their working lives through their status as shareholders.[1] As will be shown in later chapters, this is as much a failure of workers to organise collectively as it is of managerial and state machinations. At Taldym, privatisation has not instituted any form of 'workers' control', but neither has it promoted restructuring. Rather, it has helped to stabilise a particular balance of power between workers and managers in which workers have gained a limited amount of 'negative control' over management.

Taldym currently employs 2,600 workers, and until 1995 was one of the largest and most successful mines in the South Kuzbass Coal Concern. It has large coal reserves and enjoys good geological conditions. The construction of the mine began in October 1955, after a 1944 survey discovered large reserves

[1] For an account of Russian workers' failure to realise the potential for control apparently immanent in 'privatisation to the labour collective' see Clarke and Kabalina, 1995.

of coking coal.[2] In February 1956 the first mine development workers arrived in Vishnovka, and Party, trade union and Komsomol organisations were established. At this stage the Party organisation had just five members (Dolzhenok and Sal'nikov, 1982: 17). As already mentioned, the mine, then known as Vishnovskaya, produced its first tonne of coal on 1 March 1957, which is the official birthday of the mine. In 1960, the construction of a second mine, Taldymskaya, began nearby – this is now the site of the main production activity at Taldym. Then in 1970 the two mines united to form Taldymskaya mine. In November 1991 the mine privatised to become AO Taldym.

Until 1997 the mine was one of eighteen mines belonging to the concern Kuznetskugol', which was formally established in 1992 on the basis of the old South Kuzbass Production Association, Yuzhkuzbassugol'. The concern unites mines from Novokuznetsk, Mezhdurechensk and Osinniki. The main role of the concern is now to market coal produced by the mines and to distribute (steadily shrinking) state support to the mines. The mine has now left the concern – which in practice means that it markets its coal independently. The Vishnovskii section of the mine produces steam coal, although the sales from this merely cover the mine's electricity bills. Coking coal is produced on the Taldym site and this accounts for 75–80 per cent of overall production at the mine. Coking coal is sold to local metallurgical factories and processing plants. Since 1996 the mine has been making losses every month, partly because of major problems with delays in the development of new faces in the Taldym section of the mine, and partly because of the reduction in state support and investment. The mine is certainly not first in line for closure, but it is now in a far less advantageous position *vis-à-vis* other local mines than it was in the past. The main threat that Taldym faces in the near future is losing its status as a closed AO – its debt problem means that it is vulnerable to attempts by outsiders (such as the concern) to attain a stake in the mine.

The mine has always been more than a unit of production. As the Russian sociologists Vladimir Ilyin and Marina Ilyina have argued:

> The stabilisation of society through compulsory employment and minimum payment of the lowest layers of society was one of the leading principles of social policy in state monopoly socialist society. . . . The key role in the realisation of this direction of social policy was played by enterprises. It was they which provided universal employment . . . [as well as] the principal framework for the distribution of goods and services. (Ilyin and Ilyina, 1996: 382–3)

In line with such principles, employment policies at the mine were directed towards ensuring social integration of workers, rather than efficiency. Thus, the mine provided a wide range of employment possibilities to women, as well as a variety of (better paid) skilled and semi-skilled positions for men. Along

[2] A geological survey in 1856 had found no coal deposits. The more precise 1944 survey found nineteen seams, several of which were over 12 metres high (Dolzhenok and Sal'nikov, 1982: 15).

with producing coal it also ran local services. The mine has still not decisively broken with such principles, although paying for them is becoming increasingly problematic. The legacy of the past is reflected in the balance between underground and surface workers in the labour collective: almost two thirds of the mine's employees work in management and administration, auxiliary shops or *sotskul'tbyt*. Before 1996 approximately 700 of the mine's workers were women, working mainly in auxiliary collectives, such as that responsible for coal improvement, in routine office jobs, and in *sotskul'tbyt*. This number has now been reduced to less than 600, mainly because of the transfer of the kindergartens to the local authority, which resulted in the mine losing 90 employees from its books.

The mine is divided into forty-four shops, ranging in size from 180 workers to fewer than ten; the average shop will contain between 50 and 100 workers. The hierarchy in different shops varies somewhat, but generally each shop is managed by a shop chief, and below the chief (in order of status) each shop will usually have a deputy shop chief, a mechanic and several foremen and forewomen who manage the individual 'collectives' within the shop.[3] Collectives are then in turn divided into brigades, which have either an elected or appointed brigadier.

Privatisation of the mine

Taldym was among the first mines to privatise, beginning the process in November 1991. Privatisation was the culmination of the period of workers' activism at the mine which began with the 1989 miners' strike. Between the 1989 and 1991 miners' strikes the strike committee was highly influential at the mine, and it played an important role in the privatisation process. The director of the mine was pushed into moving quickly by the strike committee: he reportedly maintained that the Moscow authorities would not allow the privatisation, but the strike committee demanded action and their perception that the authorities were in no condition to resist proved correct. The closed form of privatisation in which all shares were held by employees or pensioners owed a great deal to the influence of the strike committee. According to those who were active in the strike movement, the mine was only able to secure its unique constitution because it acted quickly: 'We managed it just in time . . . while the government was still in shock' (after the miners' strikes and the 1991 coup). Managers at the mine confirmed that it was members of the strike committee who pressed for the 'closed' form of privatisation, and

[3] An auxiliary shop will typically be responsible for a range of related processes. For instance, there are three mine transport shops, all responsible for different parts of the system. The shop is then divided into collectives: thus, for example, transport shop number three includes collectives such as the coal preparation facility and the collective responsible for the repair of conveyors in the mine.

secured a constitution which in principle gave worker-shareholders control of the mine's destiny. This influence was reflected in the fact that the shareholders' council, which was established to represent shareholders, initially acted as the successor to the strike committee. As one senior manager commented:

> Here the initiative came from below. The consequences of the strike urged us on. We created an AO. Those who had been active at the time of the strike went into the Shareholders' Council. [Then] the workers' interest dwindled. 'We elected you – now pay us'. They are not interested in anything except pay.

This quotation neatly encapsulates the history of the period. Privatisation was initially seen as a major step forward for workers at the mine. Unlike most 'pioneers of privatisation' in the coal industry the mine director was not granted a controlling packet of shares to hold 'on behalf of the collective',[4] and neither were 'outsiders' given the right to obtain shares. But, as will be seen below, privatisation did not have the effect that either workers or managers anticipated.

The composition of the AO

In March 1992, shares in the mine were distributed free of charge, according to length of service (one share for every year of service). A ceiling of eighteen shares per person was set. No distinction was made between workers and pensioners who had previously worked at the mine. Workers who had been at the mine for less than a year at the time of privatisation were given the right to pay for shares. Overall 3,700 employees and pensioners became shareholders. All the shares, whether paid for or free, count as voting shares. Most of the employees at the mine are shareholders, although the number of pensioner-shareholders is obviously gradually growing.

In formal terms the highest authority of the AO is the Shareholders' Meeting, which must be convened at least once a year. All shareholders have the right to vote at meetings and their votes are weighted according to the number of shares they possess. But since, as the clerk of the Shareholders' Council put it, the mine does not have a hall big enough to hold all the shareholders, worker-shareholders elect delegates to represent them at meetings. The number of shareholders a delegate represents has varied between five and ten at different meetings. Each share counts for one vote and delegates hold a 'mandate' indicating how many shares they represent. Questions can be decided by a show of hands unless representatives of not less than 10 per cent of the shares request a secret ballot. In practice, important questions are usually decided by secret ballot.

[4] For an example of this in the mining industry see Clarke *et al.*, 1994. In the case of the mine discussed by Clarke *et al.*, also an early privatiser, the director retained a controlling interest of 33 per cent of the shares on behalf of the collective, which gave him a veto over the decisions of the Enterprise Council.

The election of delegates takes place within individual work collectives. Those elected tend to be the active and authoritative members of the collective – shop trade union presidents and brigadiers are often sent to the meeting as a matter of course. In general, collectives do not hold discussions prior to the meeting as to how their delegate should vote. A trusted member of the collective is expected to have an implicit understanding of what 'the lads' or 'the girls' think. Although collectives have some interest in hearing the results of meetings, at such gatherings, which are generally informal, it is not usual for delegates to be questioned as to how they voted on a given question.

The Shareholders' Meeting elects both the Shareholders' Council (*Sovet aktsionerov*, hereafter SA) and the mine president (as the mine director has been known since privatisation). The director is then free to choose a management team. In addition to this, according to the statutes, only the Shareholders' Meeting is competent to:

- change the constitution and issue capital;
- recall the SA;
- recall the president;
- elect and recall the audit commission;
- 'define the basic direction of the Association's activity';
- establish the yearly plan;
- distribute profits;
- create or liquidate subsidiary (*dochernye*) enterprises or affiliates;
- reorganise or liquidate the Association;
- establish the yearly budget (*Shakhterskii trud*, 4 September 1992: 1).

In practice, however, the Shareholders' Meeting does not fulfil all of these functions. It does elect and dismiss the mine president and the SA: the mine has had four presidents since the decision to privatise. It also elects the audit commission, although this body has tended to be formally judged satisfactory and re-elected every year. It has to decide on changes in the constitution, but in this case worker-shareholders are very easily 'guided' by the management team. The same is true of all the other weighty issues over which the Shareholders' Meeting apparently has the right of decision. The character of the Shareholders' Meetings will be discussed in later chapters, but here it is sufficient to note that items such as the yearly budget are prepared by the relevant manager in advance, and then agreed to by the meeting in a formal vote. Occasionally an intelligent and enquiring worker might ask a hostile question, but such interventions are not generally welcomed by the mass of the participants.

Extraordinary meetings of shareholders can be convened on the initiative of the president of the SA, at the request of the SA, the Audit Commission, or a group of shareholders holding between them over 10 per cent of the shares.

The Shareholders' Council

According to the statutes of the AO, 'in the periods between general meetings (meeting of representatives) the highest organ of the association [*obshchestvo*] is the Shareholders' Council' (*Shakhterskii trud*, 4 September 1992: 2).

Its formal duties are:

- establishment of the social and economic development plan;
- consideration and establishment of the routine budget of the association;
- monitoring the work of the administration;
- examination of collective disputes and conflicts;
- examination of applications to obtain shares;
- deciding on the issue of additional stocks or shares;
- deciding on the establishment and liquidation of joint ventures and joining and leaving enterprise associations (concerns);
- approving the candidates for the administration and concluding their contracts;
- establishment of the staffing structure in accordance with the pay fund;
- defining the mine's future direction.

The SA has eleven members, only one of whom (the president) is a permanent full-time official. The rest of the members are occasionally given time off work to fulfil their duties in accordance with the needs of the council.

Although the SA is formally the 'highest organ' of the AO, on a day-to-day basis executive power at the mine is now firmly in the hands of the mine president and his team. The clerk to the Council gave a clear assessment of the role of the SA:

> In essence it should be the owner [*khozyain*]. We privatised the mine and elected the SA which should run the mine, should conclude contracts and so on. But in reality it's not like that. Probably because we have always been used to having one boss. Now we are trying to administer things differently, but we don't know how. . . . [The president] is responsible for everything. The shareholders' council helps him, but he is the *khozyain*.

Members of the workers' movement were initially influential on the council and in the immediate aftermath of privatisation the council did act as a counterweight to the mine administration.[5] Gradually, however, as the influence of the workers' movement dwindled and workers, in the words of a former president of the strike committee, 'calmed down', the character and composition of the SA changed. From being an (albeit imperfect) representative of workers, and a thorn in the side of the director, it became just another part

[5] The strike committee itself had close links with the Council of the Labour Collective (STK) at the mine – many of the leading figures on the strike committee were members of the STK. This continuity was not broken by the establishment of the SA: now the formal and actual functions (which are two different things) of the SA and those of the former STK are very similar.

of the mine administration. The vast majority of the members of the SA are now ITR (engineering-technical workers): by 1995 only two of its eleven members were workers.

Members of the SA still claim that their role is the monitoring of the administration on behalf of the shareholders: as one of them put it, 'Our basic function is one of monitoring [*kontrol'*].' In fact, however, when the SA exerts some kind of check on the administration, it now does so in the capacity of a management faction. And generally the president of the SA and the mine president have begun to work as a team. The mine president who was voted out of office at the end of 1995 reached an accommodation with his counterpart in the SA, while the present mine director has had good relations with the SA president from the outset. The current president of the SA was Party secretary at the mine between 1980 and 1984, while the mine president is a former communist mayor of Osinniki. The two former comrades work closely together: as one worker put it, 'Our SA here is the old Party organisation of the mine.' In the partnership between these two presidents, however, the president of the SA is definitely the junior partner.

Worker-shareholders now exert virtually no influence over the work of the administration through the SA. Indeed, in common with the trade union, the most important function of the SA as far as workers are concerned is the provision of financial help. While the shop trade union committees and the trade union give out small grants to those in need, the SA is able to provide much larger sums. For example, while it was always necessary to bribe Soviet doctors in the past, workers could usually afford the sums required, or were able to secure them by borrowing from relatives. Now it is formally necessary to pay for hospital care and the sums in question are far more significant: an operation can cost several times a miner's monthly wage. In addition, many medical practitioners still expect a bribe on top of the formal payment. Since workers' savings have been wiped out by economic reform and wages are habitually paid late, workers have to rely on the enterprise to help them when they need to pay for urgent medical care. It is the SA which deals with such requests, providing either grants or loans to those in need on a discretionary basis. The 1995 collective agreement laid down that workers and pensioners should all receive the same amount – approximately $50 – in financial help every year from the SA, although in practice the individual requests for aid have continued, and the SA continues to consider them.

The mine president

The mine president is elected for a two-year term, and can be recalled before this by a majority vote of the shareholders. (This was reduced from two thirds in July 1995.) To recall a director, a Shareholders' Meeting must first vote to hold an election, and then a secret ballot is convened in which all shareholders are entitled to vote. So far no president has succeeded in being elected for a second term.

An incoming director is entitled to choose his own administration. The administration currently consists of ten posts: the chief engineer; the chief economist; the chief accountant; the chief mechanic; the deputy director for capital works; the deputy production director; the deputy director for development work (preparation of coal faces); the deputy director for social questions; the deputy marketing director; and the deputy commercial director. In his manifesto the current director promised to liquidate the mine commercial service, which no longer has a clear role now the foreign barter business has dried up, but so far he has not done so. Although every candidate for mine president promises to renew the administration and bring in new, competent people, the mine administration is actually becoming more stable. For example, although the current president was disdainful about the record of his predecessor, he still chose to keep the same chief economist and accountant in his team.

The president holds executive power at the mine. He (not surprisingly, there have been no female presidents so far) represents the enterprise to the outside world; controls the mine's bank accounts; concludes the collective agreement with the trade union; and dishes out orders (*prikazy*) which 'all workers are obliged to follow'. He is also formally obliged to implement the decisions of the Shareholders' Meeting; give a monthly report on the mine's financial position to the SA; ensure the fulfilment of the production plan (which, in formal terms, is set by the Shareholders' Meeting); and 'further the retention of qualified staff by the association' (*Shakhterskii trud*, 4 September 1992: 2). It is up to the Shareholders' Meeting to decide what to do if he fails to meet these obligations.

The impact of privatisation

Privatisation was a key plank of the neo-liberal reform package for Russia, as it was in the rest of the former Soviet bloc. But in most mines privatisation was a purely formal process because a controlling interest was retained by the state, so that neither managers nor workers had any control over the fate of their enterprise. The terms of restructuring were dictated by the state through Rosugol' and the coal concerns, which used their control of subsidies to force mines to cut costs, to divest themselves of *sotskul'tbyt* and to close what were deemed to be superfluous production units. Management responded to this growing financial pressure by delaying payment of wages, laying off workers and putting workers on short time. Workers in these mines had no channels through which to defend their interests and so, despite sporadic outbursts of protest, the predominant response of workers has been to leave the mines.[6]

[6] It is through such administrative pressure that mine closure is generally precipitated: conditions at a mine become intolerable until only those workers incapable of finding work elsewhere remain. This process was succinctly described by the author of an article protesting the 'uncivilised'

Although at the time of privatisation Taldym was a relatively prosperous mine it still depended on subsidies and so, despite its full privatisation, it was subject to the same financial pressures as other mines. The closed form of privatisation did give management more independence from the concern, and the potential to develop their own strategies. On the other hand, however, the fact that the workers held the majority of shares meant that managers had to take workers' interests into account. But workers were never able to transform their voting rights into real power, and were only able to exert a form of negative control which inhibited any restructuring perceived to be at their expense. The result was that privatisation institutionalised a particular balance of power between workers and managers at the mine, defining the confines within which management has responded to the challenges of the post-*perestroika* environment. The failure of either workers or managers to gain control of the mine has meant that privatisation has disappointed both groups.

Managers and privatisation: 'activating the human factor'

Managers at the mine had no appetite for the rapid restructuring and drastic 'downsizing' of the neo-liberal imagination, but they did hope that the mine would become more productive and prosperous as a result of privatisation. In order to understand their aspirations it is important to stress that Soviet ideas of 'efficiency' are very different from those which prevail under capitalism. Soviet managers and engineers in the past had a concern for production, rather than for profit. They were frustrated by inefficiency, waste, bottle-necks, poor quality goods and so on, but 'efficiency' in terms of ability to make a profit was not an issue: they were concerned to meet the plan and to secure the inputs they required to do so. Moreover, they also tended to be firm believers in the paternalist ethic of the traditional Soviet enterprise.[7] They saw privatisation as a means

running down of Sudzhenskaya mine in the North Kuzbass: 'the better, the most mobile workers hurry to find a job somewhere else and, as a rule, they are able to. And therefore the doomed enterprise beings to feel a lack qualified core workers. And to find them is impossible – who will go and work at a mine in the midst of its death throws?' (*Kuzbass*, 6 August 1996: 2). The trade union president at Taldym, summarising the situation in failing mines for the benefit of the mine trade union committee in August 1996, spoke in very similar terms: 'You know what happens, what we could call the most professional workers, those with connections, good workers – the people who constitute the core of the collective – they leave of their own free will and find work elsewhere. And what you're left with is an unmanageable collective [*neupravlyaemyi kollektiv*].'

[7] Even those who support the policies of neo-liberal reformers tend at the same time to adhere to traditional paternalist values. This applies, for example, to the director of the nearby mine of Abashevskaya, who is a strong supporter of 'Russia's Choice', but at the same time wants his mine to retain *sotskul'tbyt*, something which is anathema to true believers. His vision of a productive and successful mine providing a high standard of living and social protection to a hardworking workforce is very similar to that of the present director of Taldym, even though the latter is a firm supporter of the Communist Party.

to realise their desire for improvements which had been frustrated within the old system, but at the same time their aspirations were conceived firmly within the logic of the very system which they felt had constrained them in the past.

In their reflections on privatisation the managers at Taldym tend to oscillate between two visions of reform: on the one hand, the idea that privatisation, by giving worker-shareholders an interest in their enterprise, could 'activate the human factor' at the mine, and on the other, the idea that a concentration of shareholding could institute 'real ownership' at the mine, itself a dualistic concept of stewardship encompassing ideas of both discipline and paternalistic care. Thus, something very close to the dilemma of Gorbachev's *perestroika* was reproduced at the mine in the post-privatisation era: how to grant 'the people' involvement and 'an interest' while at the same time keeping them under firm managerial control.

The emancipatory vision of privatisation was based on the idea that waste and inefficiency would be reduced by the sense of common endeavour fostered by employee share-ownership. One of the traditional complaints of Soviet managers was that their enterprises were like collective farms: state ownership, like common ownership, led to universal indifference. Managers at the mine did retain some hope that once the workers felt they owned the mine they would have an interest in its success, and labour discipline and productivity would improve.[8] At the same time, however, their ideas about the evils of common ownership led them to hanker for 'real ownership' on the part not of workers but of a *khozyain*. It is again important to stress that the ideal of the *khozyain* has little in common with Western ideas about sound management, corporate governance, and still less profit. After seventy years of state ownership, the idea of the *khozyain* has acquired almost mythical properties in Russia. In the light of persistent disappointment with Soviet paternalists, many Russians have deduced that ownership is the vital added ingredient required to secure perfect performance. A real owner is a paternalist of god-like beneficence, who has an interest not only in profit, but also in stewardship: he (the *khozyain* of the popular imagination is undoubtedly masculine) looks after his workers and they in turn look after his enterprise. He is also, however, in

[8] It is true that the intensity and productivity of work at the mine have not increased as a result of privatisation. But the complaints of managers should not be taken at face value. Although they do demonstrate the negligible impact of privatisation, some of the premises on which they are based are rather dubious. It is certainly true that there are workers at the mine, especially certain brigades of male surface workers, who work at what the Russians would call 'half strength'. But at the same time, the vast majority of production and development workers work very hard, and often take major risks in order to meet plan targets (and to get bonuses), such as ignoring low levels of methane, and, in the case of development workers, using fewer roof supports than the regulations prescribe when they judge that the rock is suitable. Similarly, while it is true to say that privatisation has not transformed workers' attitude to the enterprise, managers' characterisation of workers' pre-privatisation position is one-sided. As will be shown in later chapters, workers' relationship to the enterprise is complex and contradictory: it is certainly not accurate to claim that they 'simply couldn't care less'.

control, strict, 'firm but fair': he embodies the tension between co-operative and coercive principles. Of course, while a *khozyain* might well be strict and interested in profit, the idea that he would necessarily also ensure the well-being of workers, the local community and the environment is a Soviet fantasy which has far more to do with the communist era ideal of the paternalist director than with the reality of ownership. This highlights the traditional framework within which managerial aspirations for privatisation were conceived: the fact that the passing of the communist system means that past collectivist imperatives will cease to operate is something which has only recently begun to be appreciated.

The two themes of worker involvement and the good *khozyain* can be seen clearly in managers' assessments of the impact of privatisation. What their comments also reveal is that neither of their visions even came close to being realised. First, although privatisation apparently handed ownership of the mine to the workers, no 'psychological revolution' occurred as a result of this. As one of the chief specialists lamented in a group interview in February 1993:[9]

> Nothing has changed here as a result of privatisation. Except that the president [director] has become a bit more capable – in financial matters. But the psychology of people has not changed at all – [the mine] was a *kolkhoz* [collective farm], and that's what it's remained.

Shop chiefs at the mine, interviewed in the same month, concurred with this opinion. Asked what changes had occurred at the mine as a result of privatisation they affirmed that there had been:

> Practically no changes . . .
>
> Discipline has declined, output has declined – but that is not a result of privatisation. If we are talking of privatisation, then there have been no big changes, no major [changes]. . . . People have not understood that they are the owners. And if there have been some improvements it's only those dictated by life.

Lower down the enterprise hierarchy, mine foremen, asked whether there had been any changes in the mine administration, had the same opinion:

> Absolutely none . . .
>
> There are no changes. There was euphoria, expectation. But expectations of a feeling of ownership were not fulfilled. Freedom is necessary – but that is understood as complete irresponsibility in everything. . . . Not feeling any sense of ownership, we are very frightened of civil and cultured responsibility. . . . Our tragedy is our psychology which shows itself in everything.

[9] The extracts which follow from interviews with managers in 1993 were conducted by the Russian sociologists Olga Pulyaeva and Konstantin Burnyshev (unpublished interview transcripts). Similar views have since been expressed to me by managers at all levels, but the 'focus groups' conducted by Pulyaeva and Burnyshev soon after the mine privatised produced particularly vivid responses.

Thus, managers were united in the belief that privatisation had had very little impact on the mine. It was precisely what they saw as the destructive 'collective farm' mentality that 'something that belongs to everyone belongs to no one' which the form of privatisation they had opted for preserved: as one of the chief specialists asserted, 'A closed AO *is* in essence a *kolkhoz*.' Workers, one of the shop chiefs argued, had not developed a sense of responsibility for the mine because they had been given shares *gratis*:

> Nobody knows the real value – what that share costs. The people [*narod*] have given practically nothing. . .
>
> When he [the worker] gives some kind of money for it [the share] he will already begin to try. . . . But there is nothing of that in our collective. And there won't be for the moment. Because no one has given anything as an investment in the future. That is to say, the mine was common – 'ours' – and that's how it's remained. We have not changed our consciousness. It has rightly been said that this will take decades of new generations – or a *khozyain*.

Here the other dream of managers is apparent: the idea of the *khozyain* as a universal panacea, a driving force which could substitute for the years of maturation required for the development of a 'new consciousness'. Sometimes 'real ownership' is seen as a group affair, although more commonly it is seen as something akin to Soviet-style 'one-man management' (with the magic added ingredient of the 'interest' provided by property rights). Such ideas constituted the other main theme running through the focus groups on privatisation. The chief specialists, for example, mused:

> We could get fifty people to buy up the majority of shares: . . . 'For every share I will give you ten bottles of vodka'. I'm putting it crudely, but for that money people would sell. That's what we need to do. Then the people who were actually interested in the association would manage it.
>
> Namely there should be a caucus [*kuchka*]. And to create one would not just benefit the *kuchka*, but the whole people.

The ideal of the *khozyain*, however, overshadowed that of the public-spirited *kuchka*. As another of the chief specialists incontrovertibly put it, 'I know that in a car there should only be one driver.' Another cited with approval the approach of the director at a neighbouring 'pioneer of privatisation', Vakhrusheva mine:

> At Vakhrusheva the main difference is the system of management. There one person is in control. He has got an administration and a council [of shareholders] but he decides and he takes responsibility. He immediately put the council in its place. He said 'Lads, you are good people, but what can you do? Being a good person is not a profession. You can still check up [on me]. You can measure the suit, but you can't sew it. So look, get on with your measuring and don't interfere.' That's what he did. And that's what we need to do.

In such formulations the Soviet idea of 'one-man management' clearly comes across: in this variant privatisation is seen as a means to strengthen managerial authority. As the alternately regretful and wistful tone of the above comments indicates, however, all these dreams came to nothing. No *kuchka* emerged, and the much-lamented collective form of ownership remained in place.

The structure of ownership at the mine is not, however, the main reason for the failure of privatisation. The contradictory character of the privatisation project as conceived by managers was in itself problematic,[10] but the main reason why privatisation failed to live up to expectations lay outside the mine. The managerial image of the *khozyain* implicitly situated him in a climate of plenty, a Soviet world of planned provision (only without the shortages), whereas the post-privatisation directors have actually been running the mine hand-to-mouth in a climate of uncertainty. The mine has been operating in a steadily worsening environment, and thus even had the true *khozyain* of the Soviet imagination acceded to the directorship he would have found himself unable to act in a *khozyain*-like fashion.[11] For, leaving aside the complaints about shiftless workers, one of the major disappointments of privatisation, especially for the lowest level of line managers, is that it has not given them the 'freedom' (in reality, finances) to secure the workers and supplies they feel they need. As one of the forewomen from the *kotel'naya* lamented in the summer of 1996:

> They say we're shareholders here. But it means nothing. For example, we haven't got enough fitters here. I go and ask them for more workers and they say, no, according to the norms and rules you've got enough. I say, what norms and rules? We're shareholders now, can't we decide what we need? And now the concern says we should cut 400 workers – well what sort of shareholders are we then, if they can tell us what to do? We're not. We still have the plan just as before.

Privatisation has not allowed line managers to run their shops or collectives as they would like to because the mine remains dependent on (now steadily shrinking) state support. Indeed, among senior managers at the mine privatisation is now viewed as largely irrelevant: what is seen as primary is their

[10] The contradictions involved in running an enterprise such as Taldym were well captured by a maverick member of middle management. He felt that nothing would improve at the mine until the worker understood that 'the shovel he is holding is his shovel, that he is working for himself'. But in the meantime the mine had to produce coal, and since workers had not yet understood why they should exert themselves the director had resorted to a discipline drive. This, however, in the eyes of the disillusioned mechanic, would do no good either: 'How can the boss get tough with the owners? It doesn't add up. Things will not get better through discipline.' This tension between discipline and involvement constantly crops up as an unresolved dilemma when managers discuss worker motivation.

[11] Indeed, the present director of the mine set out with very *khozyain*-like ambitions: he is a traditional paternalist, 'firm, but fair', who would have liked to institute a virtuous circle in which both mine and workers prosper. Leaving aside the intrinsic problems with this vision, the immediate barrier to its realisation was lack of money. As one of the shop chiefs succinctly put it, 'Under communism the director was more like a *khozyain*. He had something to give, so he could demand things. Now he's got nothing to give so he can't demand anything either.'

struggle for survival as subsidies are gradually reduced. As the mine's chief economist put it in 1996, 'The *khozyain* here is the state. We can't survive without the state. We spend our time trying to get money out of them, but they don't give us anything.'

This underlines the important point that what the World Bank would term 'getting property rights right' (World Bank, 1996: 6) is far from being a major determinant of an enterprise's success.[12] Formal transfer of ownership does not of itself cause assets to be transformed into capital: in the West the restructuring of unprofitable public corporations into profit-making concerns is a precondition, not a consequence, of privatisation (Clarke, 1996c: 39). In Russia most of the enterprises which were privatised 'to the labour collective' were valueless as capital: they were mainly liabilities. This is certainly the case with the overwhelming majority of deep coal mines, including Taldym, despite the fact that it belongs to the group of mines considered by the coal concerns to have a future. The mine is heavily dependent on state subsidies and privatisation has done nothing to change this.

Neither has privatisation, as the discussion above illustrates, prompted management to attempt significant internal restructuring aimed at making the mine profitable. What *is* now beginning to promote change at the mine is the alteration in the environment in which it is operating. The periodic budgetary crises which afflict the government mean that subsidies are often not paid on time and have to be extracted by strikes co-ordinated by trade unions and mine directors. Meanwhile, although the government did not formally endorse World Bank proposals for restructuring the coal industry, which initially recommended that employment be reduced in the Kuzbass coal industry by over 70 per cent by 1997 (World Bank, 1994: 44), cutting the coal subsidy is a key part of the government strategy to reduce the budget deficit.

While privatisation did nothing to encourage the sort of pruning of which neo-liberals are so fond, financial problems have led the new director to embark on a programme of staff reductions to cut production costs. In 1996 it was costing the mine 147,000 roubles to produce a tonne of coal, while the selling price was 110,000: the mine was losing 37,000 on every tonne (*Shakhterskii trud*, 7 August, 1996: 1). In response to this the director launched a modest restructuring programme which aimed to force all those working pensioners

[12] While a World Bank consultant would no doubt argue that 'getting property rights right' is precisely what the architects of the constitution of the AO at Taldym failed to do, the Bank has, none the less, begun to place less emphasis on the importance of privatisation. Although the 1996 World Bank 'World Development Report' intoned the mantra, 'decentralising ownership will be the best way to increase competition and improve performance', it placed more emphasis on financial discipline as a stimulus to reform and conceded that 'slower privatisation is viable . . . if the government or workers are themselves strong enough to keep control over enterprises and prevent managers from divesting assets, and if savings and growth in the non-state sector are high' (World Bank, 1996: 6). This admission was in part prompted by the embarrassing fact that China, which has not had the dubious benefit of a mass fast-track privatisation programme, is actually growing faster than those post-communist countries which have.

over 60 to retire; tighten up the administration of sick pay to cut down on the number of fraudulent claims; reorganise the work of the machine and electrical shops; and obtain two new complexes for the production shops and a new combine for the development shops. This was hardly restructuring on a grand scale, and the director was clear about the fact that 'without a change in government policy' his plan could not solve the mine's problems. And he has been proved right. Employment at the mine was reduced from 3,000 to 2,600 between 1996 and 1998, as mentioned above, but this has not insulated Taldym from the crisis in the funding of the coal industry. Since 1997 the mine has descended into a level of debt which means that wage delays and production stoppages are now endemic.

Workers and privatisation: 'the usual deception'

Ask workers at Taldym what they think of the privatisation of the mine and they are likely to reply that it was an '*ocherednoi obman*': 'the usual deception'. Contrary to their hopes that the privatisation of the mine might usher in a period of prosperity in which their work and worth would finally be recognised, they have in fact only experienced increasing hardship. The tide of workers' activism of 1989–91 has receded, and workers have proved unable to realise the potential for greater control over their working lives which privatisation seemed to offer. Nevertheless, although workers have not proved able to use their status as shareholders to gain control of the mine, this status has allowed them to defend their existing position: competition within management means that any director who attempts to make changes which workers perceive to run counter to their interests is very vulnerable to competitors, who can appeal to the shareholders against him. The following section will first discuss workers' assessments of privatisation, and will conclude by describing the balance of power institutionalised by the form of privatisation pursued at Taldym. The more complex question of why workers were unable to realise the apparently emancipatory potential inherent in the closed form of privatisation at Taldym is discussed in detail in later chapters, which consider the barriers to collective organisation and action among workers in the post-Soviet enterprise.

Workers' expectations of privatisation were, it must be said, rather hazy. Mainly, workers hoped that they might receive dividends and higher pay, although they did also have ill-defined hopes that they might gain a 'sense of ownership'. Workers from the technical complex, for example, said, 'We thought that things would get better, that we'd get some dividends', while those from the *kotel'naya* had hoped for:

> Something better, but everything's the same as it was before. It's probably because of the situation in the country as a whole. We thought that we'd get higher pay, but everything's the same. Perhaps if the situation in the country was better we would have noticed the change.

> *You don't feel that you are the owners of the mine?*
>
> Of course not. The mine can't even give us what we need to carry out our work – it can't give us the materials we need for the repair of our furnaces.

Privatisation has proved to be a disappointment; 'the usual deception'. Not surprisingly, given the problems of the coal industry, workers receive no dividends. Pay at the mine is actually quite high compared to other mines, and this can in part be attributed to the director's need to please his electorate. Workers, however, do not generally feel well paid, mainly because of wage delays.

They also do not feel that privatisation has given them more influence over the mine administration. One of the brigadiers from the *lampovaya*, when asked what she expected from privatisation, put it like this:

> Probably nothing. When it began there was a lot of talk. They said that the mine would be in our hands, that the collective would be able to make some decisions, but the collective can decide nothing and do nothing. Because in the administration and council of the labour collective, everywhere, there's one leadership – namely all the people who are ITR. . . . Those people are united. It is impossible to stand up against that structure. Everything is in the hands of the ITR – they decide whom to pay, whom not to pay. They hold the majority of shares; they can do what they like. The shares and votes that workers have don't count for anything – they are just for show. The simple workers don't participate in anything.

While it is not actually the case that the ITR hold the majority of the shares, it is true that, as argued above, management is still firmly in control of the mine. Even though there are competing management factions, workers play a passive role in the struggles between them: they are occasionally appealed to by the competitors, but no initiatives emerge 'from below'. For example, the SA is not a mechanism through which workers have been able to resolve the day-to-day issues which preoccupy them at the enterprise. Nor do they feel it represents their interests. A group of skilled male workers from the face machinery repair shop, for example, were very clear that the SA did not serve their needs. In their opinion it was 'just a piece of paper. They don't do anything. They just look after themselves.' The brigadier from the *lampovaya* quoted above was equally categorical: 'They do nothing. They get money and that's all. . . . A structure like the shareholders' council at the mine is not necessary. One person could do the job they do.'

Workers also feel that the SA does not function as an effective check on the administration. For example, it is clear to workers that neither they – the shareholders – nor the SA have much control over the administration's financial dealings. As one miner put it:

> We don't really participate in financial questions. We don't know what the economic service does; what's happening with our shares. . . . The Shareholders' Council is not really competent in financial questions, they can only give advice. We can only ask, where are our dividends? We're sharehold-

> ers, we privatised the mine. The economic department gives us information, [tells us] that there aren't any dividends. We haven't got any profits, or we have somewhere, but they've all been spent on machinery. And what they explain to us at meetings we don't understand: we don't have a specialised education.

Thus, there is clearly no sense among workers that the SA is a channel through which they can influence the administration: it is seen as at best incompetent and at worst self-serving.

Workers similarly claimed that the right to elect the director of the mine made no difference to them. A male worker from the *kotel'naya*, for example, was of the opinion that, 'It means nothing. We change the director every year and nothing changes.' Workers from the face machinery repair shop were equally unimpressed by 'democracy' at the mine, claiming, 'It doesn't help. They [the directors] all just grabbed what they could [*nakhapeli*].' Nevertheless, while it is certainly the case that workers have so far been disappointed in all the post-privatisation directors of the mine, they have, contrary to what the above quotations would indicate, derived some benefits from in-mine democracy.

Most importantly, workers have been given a new channel through which to express their dissatisfaction, and this has sometimes allowed them, as will be seen below, to exert an influence over questions vital to their quality of life. Privatisation has increased the potential for managerial factionalism at the mine and at any given moment there is a shadow administration waiting in the wings, ready to exploit any simmering discontent. Workers are the constituency to whom the different factions appeal, and the administration is thus careful not to pursue policies which risk uniting different groups of workers against them. Therefore, although workers constantly feel cheated by directors who do not keep their promises, privatisation has strengthened their negative control over the administration.

A good example of such control is the fact that Taldym's miners have been able to retain a shift system (*grafik*) which suits them, but which is highly unpopular with managers of both the mine and the concern. Most mines within the concern operate a six-hour shift system for underground workers. (The length of the shift does not include the time taken travelling to the face – which can be up to one and a half hours.) Managers consider the six-hour *grafik* to be the most productive; they claim that miners are unable to work effectively for a longer period. Miners, however, favour a longer shift, because, given the regulations concerning maximum monthly hours, a longer working day means fewer days a month spent down the pit. Given what has been said about the workers' 'second shift' on the land in the previous chapter, this is obviously an issue which greatly concerns them. One of the gains of the 1989 and 1991 strike waves was that some of the mines had moved over to a seven-hour shift system. This was, however, particularly unpopular with managers as it involved an hour's handover at the face, during which time, they claimed, very little work was accomplished. At Taldym, the miners were able to secure an eight-hour shift system, which has the merit of dividing neatly into three daily shifts. This meant that miners worked sixteen days a month, as opposed to twenty-three. In the summer of 1994, the concern began a campaign to restore the six-hour shift system in the South Kuzbass mines, and though there were protests, most mines eventually succumbed. Taldym was an exception.

This can be directly attributed to 'democracy' at the mine: any director who proposed a change in *grafik* would run a great risk of being ousted. One miner, asked about the benefits of privatisation, immediately mentioned the retention of the eight-hour *grafik*, which he claimed was the only gain workers had made as a result of privatisation:

> Our management want to go back to the six-hour working day, and if the mine was still in state hands, they'd have managed it long ago. . . . When we worked like that before, we were practically never at home – you'd just come in, eat, sleep, and then it would be time to go to work again. Our president [director] and all the administration are trying to persuade us to go over to the six-hour *grafik*. But now the *narod* is the *khozyain* at the mine, and we prefer the eight-hour *grafik*. The mine is meeting the plan. . . . On the eight-hour *grafik* I know that I work two days, and then I've got two days at home to rest. . . . In comparison to how we worked on the six-hour *grafik*, it's great. . . . This is the one result that we've got from privatisation: . . . an eight-hour *grafik* and we're pleased with that.

This was an assessment supported by the trade union president, and one backed up by election results at the mine. For example, in the most recent mine elections, one of the presidential candidates, a former director of the mine, stood on a platform of 'realism', which according to him entailed, among other things, a return to a six-hour *grafik*. He received virtually no support. Thus, while privatisation has not even come close to instituting any form of 'workers' control' at the mine, it has allowed them to defend their established position.

The closed form of privatisation at Taldym prevented both workers and managers from establishing firm control over the mine. In 1996 the current director confessed to being worried about the possible fall-out from his plan for the compulsory retirement of working pensioners over 60, and predicted that he might lose office before his programme was completed: 'The problem is that they [the pensioners] are the ones with the most shares. They'll get upset and might not vote for me again. You see the problem.' In June 1998 the director was still in office, although his popularity had evaporated. Wage delays had mounted and the positive side of his programme had not been realised, despite the dismissal of working pensioners. The miners still had their eight-hour *grafik* and the closed form of the Association had been retained, but workers were extremely demoralised and pessimistic about the future.

As the above quotations make clear, while privatisation did not fulfil the expectations of managers, it also failed to satisfy workers. Worker-shareholders do not feel 'a sense of ownership'; only when relating instances of negative control, such as over the *grafik*, do workers recall their status within the AO. But while workers are clear that they do not control the mine, the locus of real power is not immediately apparent. This is graphically illustrated by the confusion surrounding the identity of the *khozyain*, who is variously perceived to be the SA, the director or the state. The shifting identity of the *khozyain* is partly an expression of the objective indeterminacy in the internal and external power relations of the mine, but it is also related to the ideological conception that the *khozyain* should ensure order and protection. This means that a key feature of the alienated collectivism fostered within the labour collective of the transition period is workers' yearning for the true *khozyain* – and their continual disappointment in all the figures on whom they pin their hopes.

4

'We always lead the way': the Taldym trade union committee in transition

With the advent of *glasnost'*, the official trade unions urgently attempted to present a new face to the world. There were severe obstacles to the transformation of the unions' role at a central level, which have to date prevented them from breaking out of their subordination to the state. But were the enterprise trade unions similarly constrained? This chapter provides a case study of the attempts of the Taldym union committee to adapt to its changing environment. The analysis of the union at Taldym exemplifies the dilemmas faced by the enterprise committees of the former official unions in the transition period, and highlights the contradictions of their position after the demise of the Communist Party within enterprises. Since, as argued earlier, the trade union president at Taldym is one of the most radical and outspoken in Rosugleprof, the limits of union reform at the mine cannot simply be explained voluntaristically, by reference to his individual failings. Thus Taldym is an ideal site to examine how far the constraints on union reform are structural, and how far they are the product of ingrained behaviour and perceptions on the part of both trade union officers, the mine administration and workers.

Official trade unions after the collapse of communism

In formal terms, the official trade unions responded with alacrity to the changing political environment of the Gorbachev era. The All-Union Central Council of Trade Unions (VTsSPS) asserted its independence from the Party in 1987, and later reconfigured itself as the General Confederation of Trade Unions (VKP) in October 1990, in which the branch and republican organisations had a greater degree of autonomy. Meanwhile, from the spring of 1990, the official branch organisations began to establish republican organisations, and at this time the Federation of Independent Trade Unions of Russia (FNPR), was established, again asserting its independence of the Party and state, but also fighting for greater independence from the VKP. The branch unions initiated similar reform measures. For example, the official miners' union – which, because of competition from the NPG and the activism of the miners, was one of the most advanced of the former official unions – 'reformed' twice. First, the Soviet organisation was reconstituted as the Independent Trade Union of Workers in the Coal Mining Industry (NPRUP) at a purely bureaucratic congress that took place from 28 March to 1 April 1990 in Moscow. Then, in May 1991, the Russian, as opposed to Soviet, organisation was established.[1]

Such reforms did result in a change in rhetoric: unions became keen to emphasise their 'independence' from the Communist Party and the state, and their status as staunch defenders of working-class interests during the transition period. But this rhetorical radicalism was not matched by action. The unions' timidity was defined by their structural position as part of the post-communist establishment. Despite all the name changes and reconfigurations within the unions they retained their automatically checked-off membership, their property[2] and their control of social security within the enterprises. Meanwhile, in the early 1990s the new independent trade unions, which were experiencing major problems in establishing themselves, were clamouring for a re-registration of trade union members so that workers would be forced to choose actively to belong to a union; a redistribution of trade union property; and the removal of the official trade unions' control of social security.[3] Indeed, several abortive attempts were made by factions within the Duma and ministries to institute such changes during this period.[4] Given these pressures, the official unions were heavily dependent on the government not to cede to the demands of the independent trade unions, and thus their dependence on the Communist Party shifted to dependence on the new political authorities. Union leaders were acutely aware of this sword of Damocles above them, knowing full well that it was inertia, social security, holiday camps and the like

[1] This resulted in an elongation of the union's official acronym to RNPRUP, although it was still generally referred to as PRUP. Later, the union began to refer to itself as 'Rosugleprof' to avoid being confused with the NPG.

[2] The FNPR has substantial assets including a bank, factories, collective farms, joint ventures, sanatoria, sports clubs, hotels, houses of culture, a whole chain of printing presses, office buildings and vast financial investments (ICFTU, 1992a).

[3] For more details see Ashwin, 1994.

[4] The independent trade unions also wanted reform of the law to give explicit legal recognition to trade union pluralism and protect the rights of minority unions. In pursuit of such ends the NPG developed close relations with Yeltsin, while the independent trade union confederation Sotsprof cultivated ties with the Social Democratic party. All the independent trade unions became involved in an intense struggle with Moscow political factions to get legislation passed to remove control of social security from the official trade unions. Sotsprof, for example, directed all its energy into its alliances with the Social Democrats, who in 1992 were in control of the Labour Ministry. The NPG, meanwhile, gave strong support to President Yeltsin, and at moments of political crisis would always issue declarations pledging the miners' support to the president. All these efforts proved fruitless, however: the independent trade unions received nothing in return for their support. The fact that the government has failed to challenge the official trade unions' monopoly suggests that while these unions are dependent on the government, the government also relies on the unions to contain conflict, keeping it within the acceptable bounds of damp-squib days of protest.

which bound members to their unions.[5] So although they presented themselves as principled defenders of the working class against the rigours of reform, union leaders were in practice careful not to alienate its perpetrators.

This can be seen clearly in the nature of the protests launched by the FNPR. There is serious cause to doubt whether this organisation is capable of mobilising its members, but in any case it never attempts to do so, launching only twice-yearly 'days of action' or 'offensives' consisting of protests against the social costs of reform, at which the vast majority of participants are trade union officers, pensioners and others dependent on the largesse of the unions. The official protests of the branch trade unions (as opposed to the spontaneous actions of workers) have a similar character: they usually have the backing of enterprise directors or regional leaders who are keen to squeeze resources or concessions from the government. The nature of the protests was brilliantly captured by a critical but high-ranking trade unionist, at one time close to the leadership of the FNPR. Interviewed in April 1997, he said of the protests of the health and education unions:

> They hold strikes, but they have no impact – which means either that the strikes don't have a clear purpose or that they are just the initiative of the trade union bosses, designed to allow workers to let off steam – it doesn't change the fact they don't get paid, but it means there's not a social explosion either.
>
> I think that they work together with the regional leaders – these protests suit them – they say, 'Well done for ensuring such a peaceful protest', they keep their power and so do the trade union leaders. There are of course some enterprises where the trade unions have some initiative and have concluded good collective agreements, but very few. In general all protests are held on the initiative of the powers that be: regional governors, directors of enterprises.

This point is well illustrated by the 27 March 1997 day of action in protest at late payment of wages called by the FNPR. In several regions, including Vladivostock, Stavropol', Krasnodar, Tula, St Petersburg and Ul'yanovsk, mayors, heads of *oblast'* administration and regional governors were directly involved in the organisation of the protests along with FNPR representatives (*Rabochaya sila*, 2, 26, 1997: 2). As will be seen in the analysis of the trade union at Taldym, this co-operation with the authorities extends all the way down to enterprise level.

While union leaders' determination to hold onto their property renders them dependent on the government, this property also defines their indepen-

[5] In 1992 the (then) vice-president of the FNPR, Vladimir Kuzmenok, candidly admitted this to me during an interview conducted in my capacity as Research Officer for the ICFTU. He said that the communist trade unions had been 'a state within a state' and had very weak links with their members. He had no illusions: if the government imprisoned the whole of the FNPR leadership not one worker would lift a finger to defend them. He went on to say, however, that the trade unions were the one organisation left over from the communist period which had not been wrecked by 'so-called democracy'. Kuzmenok rightly predicted that the trade unions would never take 'Gorbachev's route' of reforming themselves out of existence (ICFTU, 1992b).

dence from their members, and as such has a corrupting effect on the trade union elite. This was emphasised by the critical trade unionist quoted above, who claimed that the trade union leaders had become part of the corrupt new elite:

> Our legislature, holders of executive power and enterprise directors, they are united – with the aim of getting as much money as they can for themselves. They informally agree to invest money that should be paid to workers in banks and get the interest – that sort of thing.
>
> In a normal country the role of the trade unions would be to monitor this, to try to keep them under control. Well, the unions create the impression that they are very unhappy about all these things, but in reality it doesn't bother them. They have been bought off. . . .
>
> Property is the main question at the central level [of the union]. Workers' interests are in about tenth place and in second place is the battle for power, the manipulation of social opinion and so on. . . . Probably there are some individuals who would like to do something. But, as we say here, 'being determines consciousness', and they operate in an environment in which they have the opportunity of grabbing property. They have no reason to do something for workers when they can get rich.

What both facilitates and reinforces this 'criminalisation and mafiaisation' of the trade union elite is their independence from their members:

> And, most important, the tens of millions of members of the FNPR don't know that they are members of this union. If you asked someone on the street if they were a member of a union they would say 'yes, probably, at least I pay dues' – but they couldn't tell you which union. They don't participate in the life of the union in any way; they don't go to meetings or anything. The trade union exists for itself [*samo po sebe*], independent of its members.
>
> They do have some kind of social influence, however, because they have offices, telephones, they can organise demonstrations when they want to, like they did on the 27th [of March, 1997].
>
> But the members have no influence over what I call the union 'bosses' – I call them that because they are like managers. At all levels the union bosses are members of the establishment. [Mikhail] Shmakov [president of the FNPR] is just a business man – he gets profit from all the trade union property.
>
> So really they are just involved in commerce – they spend their time doing business. They are basically New Russians. They are rich, they are not accountable – and meanwhile the workers are not paid. So in essence nothing has really changed in these unions – the trade union leaders are really business leaders. . . . That's why when I speak publicly I always emphasise that the trade unions and the workers live independently of each other. This is not democratic.

Such comments provide a dramatic illustration of the barriers to the reform of the trade union at the central level: there is no incentive for unions leaders to reform. To represent their members' interests during the transition would inevitably lead to conflict with the government. And to encourage a more involved membership would mean voluntarily subjecting themselves to inconvenient scrutiny.

The situation is not so bleak within the branch trade unions: the leaders of these unions do come under some pressure from officers at a lower level. Nevertheless, the 'reform' of these unions was scarcely more effective than that of the FNPR. Within Rosugleprof the reform's most important effect was to institute less authoritarian relations between the central and regional union organisations and the mine-level unions through the abandonment of 'democratic centralism'. The 'reform' of the union's structures, however, in many ways condemned it to the worst of both worlds: a top-down centralist ethos continued to dominate the union's approach to policy making, while at the same time it became unable to implement decisions taken democratically because the mine-level unions, having been liberated from democratic centralism, no longer feel themselves bound by the majority view in the union.[6] The national leadership of the union now want to reintroduce democratic centralism, but they have so far been unable to gain support for this, because one of the most important consequences of its abolition was the fact that the majority of union dues now remain at enterprise level – a situation which the mine-level unions are keen to continue. The extreme independence of the trade unions at mine level means that the question of how to reform was left up to individual mine unions.

Such developments in the internal organisation of the unions are an important determinant of their political efficacy as campaigning organisations, but such organisational issues should be separated from the main question regarding union reform: for whom is the union campaigning? Although it is theoretically within the power of the leaders of national union organisations to steer their organisations towards a 'trade union' position, 'reform' of the union requires a redefinition of the relationship between the union and its members. This can only come through a transformation of the role of the unions at an enterprise level. Moreover, given barriers to reform at a central level, the initiative for such a redefinition of relations within the unions is most likely to come 'from below'. What are the prospects of this occurring?

Former official trade unions in the enterprise: the case of Taldym

As argued in the introduction, under communism the trade unions were subordinate to both management and the Communist Party. Their main functions were to administer *sotskul'tbyt*, promote productivity and help to enforce workplace discipline. What changed after the collapse of communism? Most notably, with the ending of the Communist Party's political monopoly and its

[6] This presents a particular problem in staging strike action: without democratic centralism it is virtually impossible to maintain a solid strike. For a more detailed examination of this question, and an example of the practical problems the union faces in mounting strike action, see Borisov and Clarke, 1996.

expulsion from enterprises in 1991, the unions were liberated from their role as 'transmission belts'. Paradoxically, however, because the unions lacked the support of workers, the demise of the Party rendered them even more dependent on enterprise management than they had been in the past (Clarke *et al.*, 1993: 93–4). Lacking any other source of support, the unions were forced to rely on management to tolerate their existence. In terms of their functions, the unions ceded their formal role in promoting production and discipline, though, as will be seen below, they retained the idea that the 'general interest' was prior to the particular interests of workers (the general interest now being that of the enterprise rather than that defined by the Party). Their most important duty, however, continued to be the administration of the social services and benefits provided by the enterprise. The structural position of the unions within the enterprise has thus changed very little: social provision is still their main *raison d'être* and this is what workers expect of them. Moreover, the right to administer *sotskul'tbyt* and distribute various goods within the enterprise is granted by management and can be withdrawn from an errant union at any moment. The enterprise unions' relationship with management thus parallels that of the FNPR with the government. But does the proximity of the unions to their members at this level allow the unions to play a more independent role?

Members of the trade union committee at Taldym claimed that the union has been completely transformed since the current president was first elected president in 1991. In common with other enterprise unions, in the Soviet era the mine trade union had been subordinated to the Party. A long-standing shop trade union president was candid on this point:

> At that time the trade union just collected dues; the Party decided everything. . . . Then the Party would resolve things in favour of management; not in favour of workers, but in favour of management. A manager gave orders and that was that. . . . For example, they could say that they wanted to get rid of a certain person from the shop; just like that, get rid of them. And he would be removed, perhaps because he said something loudly about the Party somewhere . . . or said something to the manager. The manager could easily get rid of him and the trade union practically didn't intervene in such things. The only thing was, if he was a Party member the primary Party organisation in the shop might speak up for him.

This trade union officer felt that the trade union's attitude had now fundamentally changed. Asked how, he responded: 'I don't need to tell you, you can see it perfectly well for yourself. How people come and how we help them. We try to help them as much as we can. And I think this is right. Otherwise who's going to help them if we don't? They elected us to this position so that we would help them.' The trade union president, meanwhile, argued that the union had become far less hierarchical since the 1991 reform, and whereas before the mine unions had been under the thumb of the territorial committee, they were now free to organise their work in their own way. He claimed that as a result of this freedom the Taldym union committee had begun to concentrate on trade union issues and that, in contrast to his predecessors, he spent most of his time dealing with pay-related disputes; attempting to 'preserve social guarantees'; monitoring the implementation of the collective agreement and preparing for its annual renegotiation; and dealing with issues related to health and safety.

Taken at face value these accounts would suggest that the trade union, contrary to the argument of Clarke *et al.*, has been able to forge an independent path in the reform era and has reoriented itself towards trade unionism in the generally accepted sense of that term. But like most things in Russia the reality is not quite so simple. The next section will test the accuracy of the union officers' claims by analysing the union's work load. The chapter will then analyse the union's relationship with the mine administration, and consider how far the conditions promoting the union's co-operation with management still prevail. Finally, it will examine the relationship between the union and the workers, and workers' response to the union's attempts to reform.

Traditional trade unionism

Whatever the trade union president may claim about the change in trade union priorities, days spent observing events in the trade union office revealed that union officers unquestionably spend most of their time dealing with *sotskul't-byt* or related issues. The union president is deeply committed to the principle of enterprise social provision, and expends a great deal of effort attempting to preserve it. Although he commented that the Stalinist method of neutralising trade unions was to overload them with work, so as to prevent them from defending workers, he nevertheless sees 'preserving social guarantees' through involvement in social administration as a key trade union function in the transition period. Moreover, workers still feel that this is the main role of the union. The views expressed by a group of male and female workers from the *kotel'naya* were typical:

> WORKER 1: Our trade union is weak and doesn't really decide anything. You would go to the trade union if you needed medical help, if your house burnt down or something like that.
>
> WORKER 2: They don't do the work they should. They give out vouchers and financial help.
>
> *Haven't you noticed any change in the union?*
>
> WORKER 2: No. It doesn't resolve anything.
>
> MANAGER: It is, as it used to be, part of the administration.

It is clear from this quotation that the union's other activities have not yet made an impression on workers. Why this is the case is explored in more detail below, but the main reason is simply because social administration is still one of the union's chief activities.

The range of activities included in social administration is considerable. The trade union is currently responsible for controlling the housing waiting list; administering sick pay; running the *profilaktorii*; distributing vouchers for the kindergartens, and monitoring the administration of these organisations; organising the mineworkers' holidays and running the 'Hot Spring' resort; participating in the administration of DK Progress; organising the preparation and distribution of allotments; paying funeral expenses; giving small grants to workers and pensioners in financial trouble; organising the celebration of Miners' Day, and the other professional 'days'; and acquiring and distributing New Year's presents to workers. In the era 1991–4, the trade union was also responsible for the distribution of the imported 'barter' goods, such as cars, fridges, freezers and washing machines, which management acquired in return for coal. In addition to this, members of the trade union committee also participate in the social security commission (the trade union president is president of this); the pay commission; the labour conflicts committee; the health and safety committee; the *zhilishchno-bytovaya* (housing and everyday life) committee; and the food control commission, which monitors the quality of food in the mine buffet, *stolovaya* and the dining room at the *profilaktorii*. This volume of work is by no means exceptional, although the 'social sphere' is gradually contracting at most enterprises. Each mine has its own specificity – for example, at most other mines the trade union is responsible for the distribution of potato plots whereas at Taldym it is not – but generally the amount of time spent on *sotskul'tbyt* and related issues is very similar.

The first thing that should be clear from this list of traditional duties is that they take up an enormous amount of time. The trade union committee consists of fifteen elected members and has a seven-strong presidium, but only three of them work full-time for the union: the president, the vice-president and the secretary (the only woman on the trade union committee). These officials are kept very busy because many of the union's obligations are highly labour intensive. For example, the secretary spends an enormous amount of time on the paper work associated with the administration of the sick list. The president, meanwhile, is graced with a constant stream of visitors. And although some of them come with queries about their pay slips or to discuss a difficulty they are experiencing with management, the vast majority of enquiries concern social issues. The most common visitors are workers asking for financial help, or enquiring about housing, holidays, kindergarten places or allotments. The union officials also regularly receive pensioners who demand a lot of attention: the women often weep,[7] while the men are prone to deliver long speeches

[7] The trade union president is often visibly moved by such applications and usually does what he can to help. Three examples will suffice to give an idea of the types of tragedy that are regularly brought to the attention of the trade union committee. Many are housing related. One 70-year-old, for instance, came to the union after noticing a free room in her neighbouring barrack, which had

denouncing the iniquity which has led them to the door of the trade union office.

Meanwhile, in summer the union committee is laden down with the work of organising holidays. The secretary administers the waiting lists for the various resorts (at present the two main alternatives are Hot Spring – a resort deep in the Taiga and only accessible by helicopter, consisting of a cluster of huts around a spring supposed to have healing properties[8] – and a lakeside resort in Khirghizia). Every fifteen days one member of the trade union committee has to accompany a party on the twelve-hour route to Hot Spring, a task which sometimes involves spending several nights in the airport when the weather conditions are not suitable for the helicopter.[9] Meanwhile, the president has to negotiate the contracts with the other resorts; ensure that the mine administration pays the right amount, at the right time, to the holiday-makers and the trade union committee; see off holiday-makers; and, since the closure of the pioneer camp, organise a summer programme for children at the *profilaktorii.*

Another summer duty is the increasingly difficult job of organising the 'Miners' Day' celebration on 25 August. This is an event of great symbolic significance in the community, and it is important that the union gets it right. The current president has begun a tradition of organising a large concert of popular music in the main square of the settlement, but this has now become a major struggle: the most popular groups have all put their prices up and, according to the trade union president, have begun to demand 'luxe' accommodation and chauffeur-driven limousines. In August 1996 the trade union president was on the phone to music industry managers every day trying to organise a 'decent' concert, at the same time as attempting to ensure that the miners would be paid at least part of May's wages before the big day. Although he was also in the middle of planning a strike (of which more later), he seemed to be as worried, if not more so, about ensuring that Miners' Day was not a

the advantage of an inside toilet – she wanted a transfer because she could no longer manage to help her 90-year-old mother to the outside toilet of her own barrack. In a similar case, a pensioner tearfully pleaded with the union to find her a flat because her (unemployed) grandchildren, with whom she was living, were stealing her pension and eating all the food she grew. Meanwhile, financial problems are also brought to the union: one widow, for example, arrived weeping because she 'owed money to half of Vishnovka' after having given her husband what she described as 'just a decent funeral, worthy of a human being'.

[8] The trade union president had built this resort up himself. He is incredibly proud of it, and, once on the subject, can expatiate for hours on the virtues of Siberia's 'second Switzerland'. It must be said that though the resort can only cater for fifteen people at a time – the helicopter cannot carry more – those who visit it tend to come back as enthusiastic as the trade union president (if not always quite so eloquent).

[9] This is a task usually performed by the most active lay member of the trade union committee. He is a combine operator (MGVM), the highest skilled category of underground worker. He still works at the face and leads the parties to Hot Spring in his own time: he says that he gets pleasure from helping people.

flop.[10] This underlines the fact that a crucial determinant of the union's standing is its success in the organisation of such traditional rituals. Thus, the acquisition of high-quality New Year's presents is also seen as a pressing union priority. This is a situation replicated at other mines: supplying poor-quality New Year's presents has been known to lead to the downfall of senior mine managers.[11]

As discussed in Chapter 2, however, mines, like other enterprises, are gradually divesting themselves of *sotskul'tbyt*. Taldym is in fact atypical in having managed to retain so much of its social infrastructure for so long. For example, of the sixty-nine kindergartens run by mines in the Kuznetskugol' concern, forty-six had been given over to the local authorities by the summer of 1996. The concern had seven prophylactic care facilities and nine palaces of culture left on its books (Burnyshev and Pulyaeva, 1996): Taldym is unusual in having both types of facility. The divestiture of *sotskul'tbyt* does not completely deprive the unions of their 'social' role – at mines where this has occurred the unions usually continue to administer sick pay, the potato plots, the allotments, and other benefits – but it does scale it down.

This raises questions about the future of the enterprise trade unions. The end of state funding for *sotskul'tbyt* means that the unions' position as the social welfare departments of enterprises is becoming dramatically less secure and this potentially leaves many enterprise unions without a *raison d'être*. It also renders them even more dependent on management: under the communist system of state paternalism the role of the union was guaranteed, but now the form and administration of paternalist policies are a matter for the director's discretion. Even if the director does opt for a strong form of 'enterprise paternalism', the role of the union is not secure: he may choose to distribute benefits himself in order to bolster his own authority.[12] This represents a major threat

[10] The concert, much to his relief, did go ahead, but the line-up was only finalised the evening before. Unfortunately, because of the mine's limited purchasing power, it was a rather lack-lustre affair. The workers complained that it was 'boring' and they also resented the fact that 20,000 roubles was deducted from their wages to pay for the event, unlike previous years when the administration had funded it. Thus, the president's heroic efforts did not do him much good in the eyes of workers who felt that they had been yet again been cheated by the union and mine administration.

[11] This occurred at Lenina mine. In an interview in summer 1994, the trade union president gave an emotional description of the humiliation endured by his members when the administration only offered their children 'a thin slab of bitter chocolate' at new year. 'The whole town was laughing at us', he claimed. This was recorded in the mine newspaper, *Tomusinskii gornyak*, in the 14 January 1994 issue, which presented a question-and-answer session with the mine director. The trade union president concentrated on this issue and threatened that such parsimony could cause half the labour collective to leave. In the wake of this scandal the mine's chief accountant was forced to resign at a meeting organised by Rosugleprof and NPG at the mine, since one of her mistakes was deemed to be the cause of the cash-flow crisis at the time the presents were acquired. This story clearly illustrates the significance invested in traditional paternalistic gestures by both workers and union officials.

[12] Exactly this process occurred at Abashevskaya mine. There the director opted for a strong paternalist policy, but he did not choose to administer the benefits of his largesse through the

to the unions because, as already indicated above, they are partially judged on their ability to provide traditional benefits. Although the social sphere at Taldym is stronger than it is at many other mines and is still under the union's control, the union president is well aware of the danger. He has attempted to carve out a new role for the mine union committee alongside its traditional activities. The following analysis of his success in doing so highlights difficulties that all enterprise unions face in attempting to transform themselves.[13]

Trade unions and management: social partnership?

Despite the burden of social administration, the union at Taldym does also deal with issues such as pay and working conditions. The purpose of this section is to consider how far this activity constitutes a break from past practices. The first issue to be examined in this connection, before the activity itself is considered, is the union's relationship with mine management. Is the union, as the manager quoted above claimed, still part of the administration? Or is the relationship more strained and complex than this? Is union 'independence' actually possible in contemporary Russian conditions?

The immediate effect of the demise of the Communist Party was to render the enterprise trade unions even more dependent on management. However, the alliance of managers and unions was not simply an expression of craven

union. He openly said that he wanted to control the social 'rewards' on offer at the mine himself, so that he could reward the 'right' people. He completely side-lined the union, as was illustrated fairly dramatically at the collective agreement half-yearly report and amendment meeting at the end of June 1994. Throughout the meeting, proposals and questions which had a vital bearing on workers' interests were allowed to pass without comment by the Rosugleprof and the NPG organisations at the mine. Given such incompetence, at one point in the meeting the director of the mine felt the need to point out to the trade unions that he was supposed to be 'the bastard employer' and they were supposed to defend the workers. At some points there really did seem to be some confusion of roles: whilst the trade union lawyer made very few objections to proposals supported by the director, the mine lawyer was vociferous in her objection to unlawful proposals favourable to her employer! Since then the decline of the union has only continued. Although this Rosugleprof committee was particularly traditional – there was no process of 'renewal' in the union during the period 1989–91 and the same 'old communist' remained as president – it does illustrate the vulnerability of unions in the face of 'cunning and innovative' directors.

[13] *Sotskul'tbyt* is a dilemma for most mine trade unions. Most of them recognise that it is not a trade union function, but they cling on to it for two reasons. First, as argued above, giving it up would leave them without a clear role. But many of them, like the union at Taldym, have a strong emotional and ideological attachment to the ethos of social provision within the enterprise. The vice-president of the union committee at Kapital'naya mine emphasised this point. He said that if the best British trade unionist was put in his place, he (*sic*) would find that, whatever his intentions, he would end up spending most of his time dealing with the housing waiting list and the provision of buses for veterans' days out and so on. (During the interview he had received a call asking him to organise a bus for the veterans.) He said that if he did not organise such matters nobody would. In addition to this sense of duty, many trade union officers derive a high degree of satisfaction from their social work, even while they acknowledge that they should not be doing it.

dependence on the part of the unions. Just as during the communist period the labour collective had a common interest in securing maximum resources and manageable plan targets from higher authorities, so in the transition period the labour collective has a very substantial shared interest in what is essentially a fight for the survival of the enterprise. As Vladimir Ilyin has argued on the basis of research in the coal, aviation, oil and gas, and transport sectors, the 'contradictions between the interests of workers and those of the [enterprise] administration, while they are very significant, nevertheless take second place to the contradiction between the labour collective of the enterprise and the external administrative, and now even market, environment' (Ilyin, 1996b: 67). It is therefore not surprising that the former official unions 'often serve as an instrument of the struggle of labour collectives, headed by the directors, against the destructive policies of the post-Communist government which have, whether deliberately or not, led to the collapse of production, the de-industrialisation of the country, unemployment and bankruptcy' (Ilyin, 1996b: 68–9). The former president of the FNPR, Igor Klochkov, made exactly this point in an interview, saying, 'I admit that it is a bit strange to see the unions and the employers forming a common front. But don't forget that Russia is a bizarre country. . . . Today what is primary is the survival of our production. We have therefore concluded a pact with the employers' (*Le Figaro*, 27 January 1993).[14]

Before examining the attempt of the union to move into new territory, it is important to stress that, notwithstanding the trade union president's activism, such collaboration is very much a feature of political life at Taldym. This is not surprising. The mine constitutes a supplicatory unity in the face of the Coal Department of the Ministry of Fuel and Power and the government.[15] It is

[14] This comment was made at a time when the FNPR was attempting to form a political bloc with the employers at a national level. In the summer of 1992 the FNPR formed an alliance with the industrialists' organisation, the Russian Union of Industrialists and Entrepreneurs headed by Arkadii Volsky, and allied itself with the 'centrist' Civic Union. They also set up an alternative to the Russian tripartite commission, the Russian Assembly of Social Partnership, on 10 July 1992, where the employers and official trade unions were to co-operate without the interference of the government and independent trade unions. Given the weakness of the national employers' organisations, however, such alliances have not really had much influence. The most significant co-operation between unions and employers occurs at the branch and enterprise levels.

[15] The fact that until 1998 members of the mine administration still belonged to the union at the mine is one expression of this fact. This is not unique to Taldym. In 1994, the Novokuznetsk territorial committee of Rosugleprof estimated that approximately 50 per cent of mine directors were members of the union, while a survey conducted by Tat'yana Chetvernina across a range of industries found that two thirds of trade union presidents and the same number of enterprise directors think that it is normal for the managers to be a member of the same trade union, as this helps to avoid conflict (cited in Clarke, 1996b). Although there have been discussions about excluding managers from the union at a national level, at a local level it is usually argued that managers are not 'proper employers', but only 'hired workers like anyone else'. This view obviously ignores the significant ways in which the interests of managers and workers diverge, but nevertheless it does have an objective basis. The trade union president at Taldym is aware of this complexity. One day in August 1996 when I was in the trade union office, some of the mine managers phoned up drunk

fighting for a future, and the trade union president is as concerned that the labour collective survives as anyone else. This sometimes leads him to oppose the mine administration, but only within a 'one enterprise' framework.

This can be clearly illustrated through an analysis of the nature of the union's conflict with the previous mine director. As discussed in the previous chapter, the form of privatisation opted for at Taldym creates enormous potential for managerial in-fighting at the mine. This allows the union to duck and dive amid the management factions and certainly increases its influence. The trade union president is a key player in the mine's internal politics and was the previous director's harshest critic. This does not mean, however, that he sees the interests of the union and the administration as being opposed: in the case of the former director, he felt that the latter's policies were detrimental to the labour collective as a whole. That is, his criticism was conceived within a framework in which the interests of workers were identified with the interests of the enterprise. This can be seen clearly in the following denunciation of the deposed director (made while he was still in power). The trade union president felt he was leading the mine to disaster: 'He lives for the day. For him it's just an enterprise, it's not his own. What he needs, he takes and the rest doesn't worry him. If he'd worried about it, there'd have been order at the mine long ago. He doesn't want to do anything.' What the mine needed was a 'man of the people' paternalist, who fulfilled his duties as protector of the labour collective. It needed, 'someone who cares about the place – they live here and their children live here. It'll not be a person from outside, but our [person]. That's how it should be. . . . We've got to have a person in the driving seat who worries about people and the enterprise.' These comments make it abundantly clear that the president perceives the enterprise as a unity: when the interests of workers and managers diverge it is a consequence of bad management.

After the election of the new director, harmony was restored and the union returned to the familiar position of partnership with the administration. The union actively campaigned for the election of the communist director, and in turn the mine director endorsed the union president when he came up for re-election in 1996. The two began by working closely together, firmly united in their desire to keep the mine in business. This unity of purpose could be seen clearly in August 1996, when Rosugleprof was supposed to be staging a national strike in protest at the government's policy towards the coal industry. The national leadership called off the strike at the last minute, although Taldym did stage a short strike. The preparations for the strike at Taldym were marked by complete co-operation between the director and the union. The trade union president kept the mine director informed about the progress of discussions within the union, and the director gave his full backing to all the president's

from the *turbaza*. They became involved in friendly banter with the president (over the speaker phone), and said that they had just taken a decision to resign from the union *en masse*. The trade union president's response to this was: 'Go ahead, I really don't mind, it would only increase the union's authority if you lot left.'

proposals.[16] The president, in turn, put the interests of the mine first. Thus, the 'strike'[17] would not involve any workers stopping work. Instead, the mine would simply cease supply: 'we'll strike, but we'll strike in our own way', the president explained. The union also planned to blockade the railway to prevent the nearby open cast mine from profiting from the strike as it had reportedly done the last time the mine had gone on strike. At the shift meetings called to announce the strike, the president and the director firmly supported each other, both identifying the 'real enemy' as the government. Moreover, both emphasised the need for workers to 'be reasonable' even though they had not been paid for over three months. The mine director underlined the dangers of ceasing work, saying, 'We must work. . . . If we stop work we won't survive. The government will just rejoice.' The trade union president took exactly the same line at the meeting at which the trade union committee planned the strike: 'We have got to prepare for winter – if we don't get faces ready then we won't produce any coal this winter. We've got to be reasonable – not like Baidaevskaya where they flooded their own mine.'[18] (This occurred during a strike over wage delays.) He repeated the warning about Baidaevskaya mine at the shift meetings: his emphasis throughout was on the long-term interests of the mine.

This partnership between the trade union and the administration was perfectly logical, for the trade union president and the mine director were committed to ensuring the mine survived. During the build-up to the strike the director constantly recalled the fact that he was involved in a battle for the mine's future, into which he was putting all his energy. Meanwhile, the trade union president made similar impassioned speeches whenever he had an audience: as he put it at a trade union committee in August 1996, 'I live in Vishnovka and I care about it. If the mine goes down we all go down with it.' Moreover, at this time the trade union president and the director were agreed about the best means to secure survival. Both were prepared to accept some cost-cutting, but wanted to retain the paternalist ethic of the traditional Soviet enterprise. This could be seen clearly in their attitude to the 1996 Miners' Day celebration. In their

[16] At one point, when it appeared the strike was going ahead, the director visited the trade union office. 'What's happening?', he asked. 'We're going ahead with an unlimited national strike', the trade union president replied. The director smiled broadly. 'Excellent', he said.

[17] Miners' 'strikes' organised by Rosugleprof often do not involve a cessation of work, but only a refusal on the part of mines to deliver coal (which underlines the fact that managers are complicit in the organisation of strikes). This distinction is completely blurred by the Russian media, however, who report such strikes as if they were strikes in the usual sense of that term, which, perhaps not coincidentally, is useful to the government. It can eventually agree to hand out the subsidies, but only on the grounds that it is fearful of the consequences of a continuation of the deeply damaging 'strike': the whole process is political theatre designed to show other groups of workers and the international financial institutions that the government only hands out money under the pressure of extreme circumstances, in an attempt to prepare public opinion for an assault on the miners' unions. By the summer of 1998 the Russian government was showing a close interest in the experience of the British miners' strike in anticipation of an assault on the miners in the autumn.

[18] Baidaevskaya has now closed.

speeches to the shift meetings on the eve of the celebrations both of them castigated the mean-spirited moaners who said that they would rather have 20,000 roubles in their pockets than have a concert in the settlement. Both emphasised that the mine was not just an enterprise, it was a community, and if the community was to live, rather than merely survive, it had to hold on to such rituals. Otherwise, 'our children will have nothing to remember': the spirit of Vishnovka would be broken.

But such unity of purpose is not guaranteed by the fact that the labour collective has a common interest in survival. For survival in the present climate does mean cutting costs. Paternalism is only one possible management strategy – although at the moment there are still strong incentives to opt for it, in particular because directors of many privatised enterprises can be ousted by a competing management faction, sometimes supported by a vote of their labour collectives. Nevertheless, directors do not always see the maintenance of the social and welfare infrastructure as a priority, and as the financial difficulties of enterprises intensify it is common for social services and benefits to be cut to the bone, and the strategy of preserving the labour collective to be abandoned.[19] Indeed, had a former director at Taldym who ran in the last mine election been successful he would have opted for just such a strategy[20] (though the fact that he did so badly in the election underlines why paternalism is still popular among managers). Landed with such a director the union president would face a difficult choice: whether to defend the social and welfare infrastructure of the enterprise and the jobs of the less advantaged members of the labour force, or to support managerial restructuring efforts in the interests of the long-term incomes and job security of the more privileged. And even with a paternalist at the helm, such dilemmas are not entirely absent: different managerial strategies do not exist in a pure form. As discussed in the previous chapter, the present paternalist director of Taldym has already opted for some rationalisation.

Indeed, as will be discussed in more detail in the conclusion, since 1997 tension between the trade union and the director has been mounting. By the summer of 1998 the mine director had left Rosugleprof, concluded the mine collective agreement with the NPG, deprived the union of its car and attempted to divide up the trade union office, in order to remove the secretary from her en suite office to a room on another floor. The trade union vice-president explained that the director had changed: 'he wants a pocket trade union, and we won't just do what he tells us. We can't trust anything he says; we only communicate in notes.' This was not linked to any particular initiative on the director's part,

[19] Irina Kozina (1997) has argued, on the basis of longitudinal case studies of enterprises, that in 1996 enterprise managers switched from the 'social' approach to restructuring to full-blooded cost-cutting involving sacking large numbers of 'living' workers (as opposed to working pensioners, drunks and violators of discipline).

[20] He also promised during the election campaign that one of his first acts in office would be to send the trade union president back to the coal face!

nor to any radicalisation of the trade union. The director was simply finding the trade union insufficiently malleable.[21] That such a reaction was provoked without the union departing from its 'one enterprise' stance and without it attempting to mobilise the workers in opposition to the mine administration starkly reveals the vulnerability of the unions in the post-communist era.

In this sense, the union has always had an incentive to carve out a new position in the enterprise. As argued above, the decline of state-funded *sotskul'tbyt* has rendered unions far less secure. The union has reasons to look beyond the mine administration for support and to attempt to move closer to its supposed constituency, the workers. For even within the 'one enterprise' framework, the trade union's ability to invoke 'the collective' and enlist its support at key moments is an important safeguard in changing times: it gives the union limited power and a *raison d'être* beyond social administration.

The union in transition: wooing the workers

At the mines where the trade union officers are not inclined to throw in the towel, there is a sense that 'trade unionism' is the way forward. This was indicated by a comment made by several of the trade union presidents interviewed in the summer of 1994: 'we don't know what a trade union is yet; we are only just beginning to learn'. Certainly, the trade union president at Taldym is aware of the need for workers' support and does try to cultivate it. But at the same time, as the following analysis of his efforts reveals, he has also been wary not to alienate the mine administration. He is, like his colleagues at other mines, caught between pressured and no longer predictable managers and distrustful, disgruntled workers. His attempts to woo the workers are thus fraught with contradictions.

Most of the trade union president's attempts to achieve greater popularity are conceived within a traditional framework. This can be seen as a product partly of prudence, partly of habit. His heroic efforts to organise a decent Miners' Day in 1996 have already been discussed. He made a similar effort to ensure that the workers received high-quality New Year presents in 1994 and 1995:[22] in 1993 the standard of the presents had been a cause for complaint, and, as the example from Lenina mine discussed above indicates, this is a

[21] This was a striking *volte-face* on the part of the director. Before he became director he had been scornful of the NPG president. During an informal conversation in 1994, at which point he was on very good terms with the Rosugleprof committee, he had asked me what I thought of the NPG at the mine. Attempting to be diplomatic, I said, 'Well, it's very small.' 'Huh', he replied, 'It's Novikov's [the NPG president's] brain that's small.'

[22] I was present in the trade union office when the trade union president was negotiating with the supplier of the 1994 presents, and he was insistent that the quality must be very high. In 1995, the standard of the presents was so high that the chair of the audit commission at the mine attempted to prove that they had been illegally obtained!

potentially explosive issue. This attention to such paternalistic gestures is part of a wider trend. The most notable change in the union since 1991 is the increased effectiveness of its welfare provision within the framework of the paternalist enterprise of the past: the president is prepared to take the time to explain complex social security and pay regulations to workers; he developed Hot Spring so that more workers were able to go on holiday; he has refurbished the *profilaktorii*; he has attempted to prevent the transfer of the mine kindergartens to the local authority. Such activity certainly benefits workers, and does represent a change from past administrative indifference to the welfare of workers. While some members of the trade union committee see this development as evidence of genuine reform in the union, however, it still stems from a traditional conception of the unified, paternalistic community of the labour collective, which constitutes workers as individual supplicants. So while it does something to improve the standing of the union, it does nothing to transform the nature of workers' relationship to either the union or their employer.

Alongside this very traditional approach, however, the union has also attempted to become more responsive to workers' complaints. For example, as mentioned in the previous chapter, it supported workers against the administration over the issue of the *grafik*, and generally the union will support workers if they begin to mobilise independently.[23] This does represent an improvement: the union would only support either the Party or the mine administration in the past. At the same time, however, it can also be seen as an attempt on the union's part to prove its utility to the administration as much as to workers: by negotiating a compromise the union can actually keep conflict under control. Moreover, on other day-to-day issues, where there was little potential for independent mobilisation, the union, like its counterparts at other mines, was far less active.[24]

The union committee wishes to avoid such conflict because its precarious position obliges it to maintain a delicate balance between the interests of the mine administration and those of workers. The union president negotiates this contradiction through the 'one enterprise' ideal that a virtuous circle can be created in which improvements in the position of workers lead to higher productivity and hence benefit the enterprise as a whole. He does try to resolve long-standing grievances, but at the same time he does not promote the mobilisation of workers: this could initiate conflict within the collective that the trade union would be unable to contain. For it is in the interests of the union to

[23] This was also the case at other mines where the union committee was quite active. At Baidaevskaya mine, for example, the mine trade union also supported the miners' attempts to retain the seven-hour *grafik*. At Abashevskaya mine, where, as discussed above, the union had been side-lined, the director returned to the six-hour *grafik* without any difficulty.

[24] The trade union's action over the *grafik* did not do it much good in the eyes of workers: only the wife of the trade union vice-president spontaneously cited this as evidence that the trade union had changed.

prevent any conflict reaching the point where it would be faced with a stark choice between siding with the workers or supporting the administration. Where the issue is not too controversial, as in the case of the *grafik*, the union president can pose as an 'honest broker', but if the conflict is more serious it threatens to expose the contradictions of the union's position. This desire to meet some of the workers' demands without shattering the 'unity' of the labour collective can be seen clearly in the union president's approach to the controversial issue of the mine's payment system.

In the summer of 1995 a struggle began between the trade union and the previous mine president over the issue of the payment system. The twists and turns of this story, which is related in full below, provide a graphic illustration of the way the trade union president is attempting to deal with the contradictions of his position, as well as highlighting the limitations of his political approach. The struggle began over the union's attempt to reform the pay system: the perceived irregularity and injustice of the pay system were and are a major focus of conflict at the mine. The trade union president identified the main problem as uncertainty: workers were never sure how much they were going to get at the end of the month, so there was no incentive for them to work. This was because of the interference of the mine director, who, according to the trade union president, gave decent wages 'to those he likes, to "his" people'. Also, while the work of the underground development and production workers was nominally tied to their fulfilment of the plan, the pay of the surface workers was dependent on the mine's fulfilment of the plan and not on their own level of work. The trade union president wanted to introduce a new system to remedy these errors. Under the system that he sought to introduce, he claimed:

> The most important thing is that people know what they are going to be paid at the beginning of the month – they can clearly see what they are going to be paid if they fulfil the plan. And the new system can't be interfered with – it is a clear formula that can't be changed. So it will be independent of the mine president, he won't be able to meddle any more . . .
>
> The new system will improve discipline. It means stability. As I said, if a person knows what he's going to get at the end of the month, he will have an incentive to work. The essence of the new system is that the approach to *everyone* is the same. Not like now when the approach to everyone is different, depending on their relationship with the director. There are also productivity inducements in the new system because it doesn't just depend on the fulfilment of the plan, but the number of workers that it took to fulfil the plan – so there is an incentive to keep numbers down . . .
>
> People must have an interest. The work of builders must depend on how much material they produce. Pay must depend not on coal production, but on [a worker's] own work.

The new system that the trade union president was seeking to introduce was one which addressed the grievances of workers, who had long complained that they were unable to understand the basis on which their wages fluctuated and that there was favouritism at the mine. What the above quotation makes clear, however, is that the trade union president was not only interested in increased fairness, he was also interested in discipline, stability and productivity: his plan was conceived within the framework of the 'one enterprise' vision.

At the same time, however, there was a strong populist thrust to the trade union president's proposals. His preferred system would reduce the pay of the (hated and despised) lesser administrators and clerks (though not of the more powerful members of management):

> The pay of the chief specialists – the chief economist, the chief accountant – won't fall. But the middle-level personnel, simple clerks, who don't influence production in any way, simple accountancy clerks and book keepers – basically the book keepers – they have always been a *bête noire* [*bel'mo v glazu*] here. Their work load does not alter whether the mine fulfils the plan or not. They write their little bits of paper, just as they used to write them. But they get paid like core [workers]. . . . And they don't need to fulfil the plan. They don't need to do anything.

This bias exactly reflects the opinions of both male and female workers, who, though they often criticise top management for being incompetent, reserve their most bitter disdain for the 'simple clerks', whom they view as morally suspect parasites who contribute nothing of value to the mine.[25] Moreover, there is a particular animus against female administrators, who are viewed as self-seeking and disdainful. Their *hauteur* is a major cause for complaint, particularly since it offends against workers' ideas about the dignity of labour. For example, an almost universal complaint of women workers was that women in the pay department were consistently rude and refused to answer any queries from workers. Thus, the proposed scheme of the trade union president was likely to appeal to workers – it improved their wages at the expense of the most unpopular employees at the mine.

The trade union president therefore appeared to have a very winnable case. He did not, however, seek to enlist the support of workers for his scheme. The scheme was due to be discussed at the meeting convened to consider and sign the new collective agreement at the end of June 1995. But the trade union president did not distribute any literature about his proposal, and neither did he instruct the shop trade union presidents to discuss the issue with workers in their shops. In fact, he did nothing to prepare the opinion of workers, or to mobilise them behind his scheme.

This showed at the meeting. The collective agreement was concluded first, and then the meeting went on to discuss the proposed change to the pay system. The chief specialists all spoke against the proposal, arguing both that the mine

[25] During the 1989 strike demands for reductions in administrative staff commanded universal support among miners (Clarke *et al.*, 1995: 39). This hostility has not diminished: most candidates for the post of mine director at Taldym have the promise to reduce administrative staff as part of their programme (though none of them acts on it once in office).

could not afford it and that the pay of some workers would actually fall under the new system. When it came to a vote the proposal was defeated. As the trade union president admitted, the workers were afraid of the consequences of the new system:

> They were scared. The specialists tried to scare the workers, saying that they wouldn't get a premium under the new system and so on. People are always scared of the new. They are used to the present system. For example, the face workers were scared because the new system takes into account productivity. So those whose bosses invite a lot of *blatnye* workers [workers who are given jobs as a personal favour] into the shop were scared that this would affect their pay. They were scared that because of what we call the *podsnezhniki* ['snow-drops' – workers who do not actually perform the job for which they are supposedly paid] they would get lower wages.[26]
>
> But this is not really a problem. Under the new system everything will be clear. Workers will know how many workers there are in the shop and will understand the basis on which their wages are calculated. They will demand that their bosses reduce the number of *podsnezhniki*. They'll say, 'Excuse me, my respected one, I'm not going to feed your "children", I've got my own to feed.'

Whatever the real implications of the proposal, however, it seemed that, through his failure to prepare and mobilise workers behind his reform, the trade union president had lost out to the presidential faction of the mine administration.

But this was only how it appeared. In fact, the trade union president had been far more cunning than this. He had slipped a clause into the collective agreement saying that a new pay system had to be agreed by the trade union and management by 1 July. This meant that there would be a legally binding obligation to introduce a new system, and he was confident that his proposal would be accepted: the new system had to be agreed by the trade union and he would not put his signature to any other. As he gleefully explained:

> The most important thing . . . was that I got the point in the collective agreement, and that it was agreed. In that point it says that the pay system will be agreed at a meeting, and will be changed only with the agreement of both sides. I won what I needed to. . . . The discussion which went on after the acceptance of that point in the collective agreement already didn't worry me. It was already a victory . . .
>
> They [the specialists] didn't pay attention. They thought, OK, let that point be . . . I didn't argue with them strongly so that everything would be OK with that point . . . I immediately at the conference proposed that he [the director] sign the collective agreement, I signed it myself and put it in the archives. That's politics . . . you work out your every step. I knew that if I introduced the pay system straight away, the collective agreement wouldn't be accepted. . . .

[26] There was a lot of heated discussion about *podsnezhniki* at the meeting of the labour collective – the administration did not attempt to deny their existence.

> That's why I separated the collective agreement and the pay system. A cunning game.

The president had been Machiavellian down to the last detail. The way that he had proceeded meant that workers would get their first pay packet under the new system in July, rather than June, and, as he explained, 'June didn't suit me':

> In July the mine will fulfil the plan and there'll be a sharp increase in pay – and that's a solid indicator for the new system. Then no one will drive people back [to the old system]. Good people are quick to understand. If it had been introduced in June . . . we didn't fulfil the plan and there would have been a slight fall [in the wages of] the surface workers. There would have been opposition – you can't explain anything to people.

His apparent defeat was thus actually part of the skilfully executed plan of a political virtuoso.

But the story did not end here. The point was in the collective agreement, and the president thought that he would be able to push through his system, claiming that he would not put his signature to any other. He was, however, unable to do so. The administration simply refused to accept his system, on the grounds that they could not afford it. Instead they introduced a different system, which retained the main fault of the old system: it is, according to the trade union president, 'slippery; the administration can manipulate it to further their own interests'. Nevertheless, the president gave his formal agreement to the new system, and the boast that he would take management to court if they did not fulfil their obligations thus came to nothing. This clearly illustrates the fact that in the last instance the president will not oppose the administration.

What it also highlights is the limitations of his approach: as an individual he is obviously weaker than the mine administration. The trade union president conceded that he had been defeated: 'they were well prepared. They are strong. I'm obviously weaker than them.' Had he, however, decided to involve workers, rather than fighting a one-man campaign, the administration might not have been able to defeat his proposal so easily; because, as argued above, had he explained his plan to workers, it is likely that they would have been behind it. As it was, they were not even aware that the point had been slipped into the collective agreement and were therefore not surprised when the pay system was not reformed.

The president, however, did not want to involve workers. There are two reasons for this, which both relate to the nature of transition. First, objectively it was far easier for the union president to attempt to gain his point through cunning politics: his approach meant that he was in complete control of the process. He did not have to deal with all the contradictory demands which could have arisen from different collectives had he chosen to discuss it with them. He could present a neat package without having to go through a process of complex collective deliberation. Second, it is clear that it would be inconvenient for the trade union to be required to deal with workers as a mobilised collectivity: this could disrupt his relations with management, as well as causing tension within the labour collective. As a passive mass, workers can be wheeled on and off the political stage as required, but attempts to involve them risk sparking off conflicts which cannot be resolved within a 'one enterprise' framework.

The union president did not, however, necessarily conceive this at a conscious level and certainly did not confess that this was his aim. Indeed, he frequently laments the passivity of workers. But his reflex action is to attempt to dampen down conflict and avoid mobilising workers. This is not only a calculated reaction to circumstance, but also a product of history, as can be seen in the following comment on the passivity of workers: 'There are lads with initiative, but very few. Generally they keep things among themselves. To come forward and demand things – it doesn't happen. . . . A Russian person is very patient. But when he breaks out – then it's a nightmare.' This view of the destructive potential of Russian workers is common among union officers and bears the clear imprint of the past – not only the Leninist horror of any 'spontaneity' among workers, but also the intelligentsia's fear of the *bunt* (mindless riot), which goes back to the Tsarist era. Thus, while there are objective reasons why the union might want to suppress the expression of contradictions within the labour collective, it also comes 'naturally' to do so. This reflects too an ingrained form of politics in which production is primary and the mobilisation of workers is actively discouraged: in the past any form of collective action on the part of workers was the enterprise trade union president's worst nightmare.

Another way of viewing the union president's approach would be to argue that it is not so different from that of union negotiators in a country like Britain. Writers such as Hyman, for example, have commented on the way that the use of specialist negotiators distances trade union members from the union and engenders a politics of accommodation: 'collective bargaining undertaken by "specialist" negotiators *on behalf of* the broader membership consolidates a representative hierarchy functionally oriented towards accommodation and compromise with capital and its agents. . . . Representation is detached from mobilisation' (Hyman, 1989: 181). The Taldym trade union president's 'specialist' approach to the introduction of the new pay system could be seen as merely an example of this type of politics: trade union negotiators in Western trade unions may sometimes also want to avoid having a mobilised membership on their backs. But it is more complex than this – as argued in the introduction, Soviet trade unions were never constituted to represent workers' interests. The politics of the Taldym trade union president are the product of *transition*: the union is caught between the past, represented by the alliance with management and fear of the workers, and an uncertain future in which its new relationship with workers is yet to be defined.

Looking more widely than the issue of the pay system, there are clear structural constraints – both subjective and objective – on the action of the union officers in most areas where reform is needed. These extend beyond the limits of the union's relationship with management both upwards, to include the terms on which the enterprise is integrated into national economic structures, and downwards, to include the subjective expectations of workers. This is starkly revealed by the union's approach to what its president referred to as the 'irresolvable' problem of health and safety violations at the mine. These are encouraged by the fact that miners' wages are still determined by plan fulfilment rather than on a time basis. Thus, as in the past, production and, especially, development workers regularly risk their lives for the sake of a few extra roubles. Workers often complain about the poor organisation of work, but they do not consistently oppose the system, and usually break the regulations on their own initiative. All parties tacitly collude in the system: in place of reform the relevant managers are sometimes sacked after particularly tragic incidents in a strikingly ineffective form of collective contrition. The trade union president's comments on this issue were highly revealing:

> There are necessarily violations of the safety regulations in order to meet the plan. There are not enough materials to allow work to be carried out according to the regulations. . . . Usually an accident is the fault of the administration. . . . Several shop chiefs have been sacked for their role in accidents. Once the chief of the technical department was sacked after an accident. But all the same accidents continue to happen. . . . The indestructible system causes them. The government can't prevent accidents. Neither can the government inspectors. These are not independent organisations – someone can always make a phone call to the relevant person to sort things out. The problem is that *there are no independent organisations which could prevent these accidents* [his emphasis]. They do not exist here.

The trade union president seemed quite oblivious to the fact that he was the leader of the supposedly 'independent organisation' which in what Russians would term 'normal countries' would be responsible for playing this watchdog role. But it is not simply a question of ignorance. Workers are not pushing him to improve health and safety: they are far more concerned, as will be seen in later chapters, about issues of distribution. Meanwhile, the trade union is operating in a situation in which the law is habitually ignored and every official has his or her price: no one is *expected* to take a principled stance about violations of laws and regulations and in subjective terms to do so is almost 'unthinkable' in the literal sense of the word. Instead, the traditional Soviet obsession with punishing *individuals* for the manifestations of systemic deficiencies prevails, and nothing changes.[27] Given such objective and subjective constraints it is easy to

[27] For a brilliant analysis of the quintessentially Soviet devotion to individual punishment see Kiblitskaya, 1995. In systemic terms this practice serves to prevent structural change and protect vested interests: a manager or two is sacrificed but the system remains. Meanwhile, in subjective terms the obsession with individual guilt goes hand in hand with the idea that having a good leader or good *khozyain* is the cure for all ills.

see why Western trade union training courses have such a limited impact in the Russian context.[28]

Workers and the union: scepticism and distrust

Despite the apparent activism of its president, workers at the mine are cynical about the union. For example, workers would often ask me why I had chosen to do research at Taldym. If I responded that it was because their union was one of the most active in the area, they would snort in disbelief: 'It's no different from any of the others.' It is, however, important to note that workers have a deeply ingrained distrust of authority and denunciations come easily to their lips: as the trade union president complained, in Russia anyone who became an authority figure – anyone who 'stood up in front of the microphone' – immediately became an enemy of the people. Certain complaints have a ritual quality, and they may not accurately reflect the opinions of those who voice them. A common complaint, for instance, is that members of the mine trade union committee are only concerned with lining their pockets and looking after their own; there is a housing shortage, but the trade union committee all have somewhere to live.[29] In a burst of enthusiastic denunciation one woman worker suggested that members of the trade union committee had several places of residence! On further acquaintance, however, it turned out that this woman had worked with the respective wives of the trade union president and of his deputy for years, had good relations with both of them and was well aware that they did not have 'several houses'. The same point is illustrated by this miner's remarks about the trade union:

> Our trade union – we need to wring things out of it, beat things out of this trade union. But they look after themselves all right. They'll do anything for themselves. . . . For example, the mine has its own farm. They write themselves out some kind of financial help,[30] take a pig . . . and off they go – the trade union live it up at the *turbaza.* Make themselves *shashlik* . . .

[28] The trade union president at Taldym attended a training course in Germany in 1994 and came back very impressed with the trade unions there. He attributed their success to their legislative base. But Russian trade unions are actually in an enviable position with regard to the law. Their problem is that they are unable to secure its implementation.

[29] The trade union president and vice-president do live in the best accommodation on offer at the mine, although their secretary, with whom the president is on very good terms, lives in a cramped two-roomed flat with her husband and two sons. The president studiously ignores her 'joking' pleas: 'couldn't you just find me a nice new flat? No one would find out.'

[30] The same miner reported in a later interview that on a visit to the trade union office he had witnessed the trade union vice-president say to another member of the trade union committee, 'Here, fill out a request for funeral expenses – we need the money for the celebration of the trade union day at the *turbaza.*' The union does sometimes write out receipts for fictitious funeral expenses when it needs money, although on the occasion when I witnessed the president do so, it was to help a pensioner. An old man arrived, needing money to celebrate his seventieth birthday.

> Now there's one fellow there – the president . . . He approaches these questions humanely [*chelovechno*]. He explains things to anyone who goes to him, tries to show them, help them. . . .
>
> There's one man in the trade union committee who'll fight for our rights and that's the president. He'll at least explain things to the men.

Workers' initial responses to questions about the effectiveness of the trade union are thus at least partly a reflex action. They certainly resent the trade union, but at the same time there is some recognition that the current president represents an improvement on his predecessors. Although it is partly a hangover from the past, however, workers' distrust is none the less potent for that: although they may modify their denunciations on reflection, workers' action in concrete situations will often be swayed more by the weight of experience than by considered reflection.

Another difficulty with which the union has to contend is the fact that workers' expectations of it are rather contradictory. On the one hand, they say that the union does not do the work that it should and only performs social work, and on the other, as seen above, the workers do actually assess the union officers on their ability to deliver social welfare. Thus, any union that attempted to become a 'pure' trade union, unencumbered by the duty of social provision, would risk losing its members. For one of the advantages of the union's present position is that it gives workers a reason to remain in the union, a fact of which the trade union president is fond of reminding anyone who will listen: 'These people leave the union, and then someone dies, or something happens, and then they come to me for help, and I say, "You should have thought of that".'[31] This was a problem encountered by the independent trade unions: they were expected to match the benefits of their official rivals. Thus, the union is partly constrained by workers' expectations. Asked what a union *should* do workers will, almost without exception, reply 'defend workers' interests', but they also expect a whole range of services alongside this. Moreover, in practice many workers would not support a union which acted to 'defend' them by taking a principled stance on issues such as health and safety or favouritism. As will be seen in later chapters, though workers complain bitterly about the social organisation of the post-Soviet enterprise, they themselves are implicated in many of its practices.

He struck up a lively conversation with the president about Yeltsin's deficiencies and the two obviously warmed to each other. Explaining to his vice-president why he had decided to give the man money, the president said, 'He was genuine, he's never been in here before asking for money; I thought, let him have a nice birthday.'

[31] I witnessed a worker attempting to leave the union, because, he said, he had never been on holiday. The secretary smiled, and simply reminded him that if he left he would never receive help from the union again, help which he would certainly need when he retired. As the NPG president periodically reminds workers through the mine newspaper (he is friends with the editor), non-union members are entitled to receive sick pay and other state benefits administered by the former official union. Nevertheless, workers still stand to lose the discretionary help available from the union if they leave.

In terms of workers' assessment of the union's performance in social administration, the Taldym trade union committee does not get a good return on its investment of time. There is a recognition among some workers that, notwithstanding the decline in enterprise social provision which is afflicting the whole of Russia, the union's approach to social administration is fairer than it was in the past. As one miner commented: 'It is easier to go on holiday now. Before I couldn't get anything out of this trade union. You would go along to the office and they would say there is no chance of you going on holiday and that would be that. The trade union is more approachable now. It will try to help you.' But not all workers would agree with this assessment. Indeed, the above comment was made in a group interview of miners, and the participants were completely divided as to whether the union had improved since the election of the current president. Thus, although the trade union president expends an enormous amount of effort on maintaining and improving social services, this has not had a decisive effect on his standing. This situation is only likely to be exacerbated as the crisis of funding in the social sphere intensifies.

The union's attempts to move into the realm of 'defending workers' interests' have also made a limited impression on workers. The description of the president's campaign to improve the pay system provides ample explanation of why the benevolence of the trade union often falls on such stony ground. The initiatives of the union do just descend from above: when the union does 'win' something for the workers, they are often unaware of the fact. This could be seen in the negotiation of the 1994 collective agreement. In a group discussion with a mixed shop of development and face workers in the summer of that year, it emerged that a major cause for grievance was the fact that face workers were paid more than development workers. The trade union president, who was eavesdropping, interjected that in fact this problem had been remedied by the recent collective agreement and they would notice this when they received their next pay packet. The miners knew nothing about this: none of them was aware of what the union had negotiated on their behalf. Therefore, although the union was obviously sufficiently in touch to pick up on issues which concerned the workers, its action did nothing to alter the relationship of the union and its members. But for this conversation, on the receipt of their next pay packet the workers might well have attributed the change to the good will of management. This again relates back to the union's contradictory position: if it advertises plans to remedy problems at the mine it risks promoting worker mobilisation around the issue concerned and angering mine management by its 'interference', but if it works quietly behind the scenes, doing what it can to remedy the worst abuses, its relationship with workers will not change, and it will have no one to defend it if management decides to jettison 'social partnership'. The union's equivocal stance thus does inhibit the development of a new relationship with its members, and provides justification for another of the workers' traditional complaints that, notwithstanding its apparent activism, the union is 'just a pocket trade union'.

This chapter has shown that there are structural constraints which limit the potential for union reform: for example, it is not easy for the unions simply to abandon social provision. The continuation and reformulation of past practices (such as paternalist provision) reflect not only external constraints but also the internalised conceptions of union officers and workers. At present, the external environment has not changed so much as to render the concepts of the past invalid, or necessarily to make alternative courses of action feasible. But significant changes are occurring: *sotskul'tbyt* is no longer state-funded and subsidies are being cut. Management is facing tough choices, and so, in turn, are the unions: the trade unions will not be able to maintain their 'one enterprise' stance forever. The union at Taldym is at least attempting to move closer to the workers, although it is still far from clear which way it would jump in the event of a more stringent restructuring programme at mine level. At the majority of enterprises the unions have not even gone this far and many conservative unions, like the one at Abashevskaya, are effectively already dead. Where, as at Taldym, the union officers are determined that their organisations survive, the crucial question is how the unions' relations with workers will develop. There is a chance that the present alienation of the trade union president from the mine director may lead him to move in the direction of involving workers, but it is just as likely to lead him to forge alliances with opponents to the current director within the mine administration, as he did during the term of the previous director. As of June 1998 there was no sign of a change in direction; the trade union president appeared instead to be slipping into despondency.

This analysis raises important questions about the relations between unions and the workers. To drop their equivocal stance in an all-out attempt to gain the trust of their members would be an enormous risk for the unions, especially since the reaction of workers to even such full-blooded advances is difficult to predict. But is there no prospect of 'reform from below'? Could the activity of workers, at a time of increasing conflict, not initiate the transformation of the unions? The next chapter will begin to examine this question, through an analysis of the shop trade union committees at Taldym.

5

Shop-floor trade unionism: the prospects of 'reform from below'

If there is a chance of 'reform from below', it is most likely to come from the shop floor. Therefore, the logical place to look for the beginnings of such a process is at the level of the shop trade union committees, which are the point of intersection between the union and the workers. As argued in the previous chapter, at the mine level the union still operates in partnership with management, albeit on a more precarious basis than in the past. But do the shop trade union committees operate within the same constraints as the mine union committee? Tension in the mining industry is mounting and spontaneous strikes are occurring increasingly frequently. What impact is this having on the shop trade union structures? Are they playing any role in channelling the conflict, or is it passing them by?

The fact that the unions do have shop-level structures is obviously an advantage. It is important to note from the outset, however, that in the past these organisations had a very low status within the enterprise, and were firmly subordinate to the shop Party committees. As one of the most active shop trade union presidents admitted when asked about the division of labour between the shop trade union committee and the shop Party committee in the past, 'If I'm honest the trade union probably listened to the Party. . . . The Party was the head of the family at shop level.' Given this history of subordination, it is not surprising that at many mines the shop trade union committees are not a picture of institutional health. This, for example, was the assessment of the newspaper of the South Kuzbass Rosugleprof territorial committee, *Gornatskaya solidarnost'*, made in an article by V. Nachalova on shop trade union committees published on 30 September 1994:

> There are a lot of presidents of shop committees who don't want to work and don't deserve the trust of those who elected them. It is a paradox that in many collectives workers don't even know who is the president of their trade union committee. It's dreadful that one encounters such facts among workers from core professions where every day miners face a whole range of problems. And there are so many of them! The frequent violation of safety regulations, the weak production culture at work, the late payment of wages . . . the collective should resolve such problems together.

The author's explanation for the weaknesses she identified was that union elections were taken too lightly; most workers did not care who was elected to their shop trade union committees. This line of argument continues a long tradition of ritual Soviet breast beating: it does not explain anything. It blames the weakness of the shop trade union committees on the fact that workers do not take them seriously, but it fails to ask why the union elections are of so little consequence to ordinary workers. In fact, however, this question is the crucial one: how do workers relate to the shop trade union committees, and if they do not take them seriously why is this? Again, Taldym is a good place to examine this question because its shop trade union committees have not atrophied in the face of change, and exist in all but the smallest of the mine's forty-four shops. Although they are habitually referred to as 'committees', generally all the work is performed by the president and vice-president of the shop organisations.

The role of shop trade union committees

In order to assess whether the shop trade union committees are amenable to reform initiated by workers, their role within the enterprise must be examined in detail. According to the trade union president at Taldym, in the present period the main role of the shop trade union committees continues to be social administration of the mine: they administer the 'sick list' at shop level; they deal with issues relating to accidents at work and at home; they distribute goods in short supply and certain privileges, while the shop trade union presidents are supposed to be the first port of call for workers with queries about holiday vouchers, social guarantees or housing. Meanwhile, each shop committee also has its own small budget for financial assistance to which workers can apply. As far as their 'representational' and political duties are concerned, the shop trade union presidents organise the election of delegates to mine trade union conferences and they are required to report to members of their shops about the results of such conferences as well as about other trade union activities. In formal terms, the unions are also supposed to represent the workers in their shops, but in practice this is the least important part of their work.

Like the mine trade union committee, the shop trade union committees spend the vast majority of their time on 'social work'. This involves duties such as distributing goods to workers that the administration makes available through the trade union at reduced prices, and providing small amounts of material help to members.[1] This face-worker's assessment of the role of the shop trade union committees was typical: 'They can provide financial help from their fund. They can provide vouchers for visits to the *profilaktorii*. The trade

[1] At some mines the union is responsible for all such distribution. At Taldym, however, there are two parallel distribution networks. The trade union distributes some goods, but the mine also has a commercial service which makes mainly imported goods in short supply available for workers to buy through a direct deduction from their wages (*pod zarplatu*).

union now distributes barter, goods. That is, it works out that they deal with material help and things.'

The shop trade union presidents themselves provided a similar assessment of their role, again claiming that their main tasks were distribution of goods and the provision of financial help. This, for example, was the assessment of the female president of the predominantly male ventilation and technical safety shop:

> Most of the workers in my shop are men and I'm like a second mother to them. I get them products, give them vouchers for hot food, they come to me if they have an accident at work. They come to me like to a second mother, like I said. It's my third year, and I think they're all satisfied with me. When it comes to dividing up goods I do everything properly. I have an exercise book where I write everything down. I do it fairly, I like to help people. . . . In the shop trade union committee we now have our own budget for financial assistance. Lots of people come to me for help.

Her male colleague in the electrical mechanical service did not contradict this. Asked to list his main duties he replied:

> I help people; give them financial assistance when they need it. We distribute goods – cars especially. We divide them according to factors such as length of service. We give out information – which vouchers people are entitled to for holidays, which privileges they are entitled to. . . . We also organise the workers' holidays, and the distribution of the New Year's presents.

When asked about the sorts of issue with which workers were most likely to approach him he answered:

> Most often they need some kind of help – financial help. Often it's the director's responsibility to look after their needs, and in this case we help them approach him. People are different – some people get embarrassed about going to the director for help, so we help them; ease their path, help them get what they need out of him.

Thus, the shop trade union committees above all look after workers' social needs through administering the social, and part of the commercial, services provided to workers by the enterprise.

This certainly gives the union a hold over workers, but, as the previous chapter revealed, the administration's delegation of these distributive tasks to the trade union is something of a poisoned chalice. Neither the social nor the commercial services run by the union can meet the level of demand, and this is by no means a situation confined to Taldym: indeed since the mine is one of the more successful in the Kuzbass, its levels of provision are superior to those at most other mines. In this shortage economy, conflicts over the justice of distribution are endemic. The availability of a limited amount of highly sought-after goods at favourable prices, for example, not surprisingly causes a good deal of trouble and the opprobrium thereby generated attaches itself to the trade union committees and does nothing to improve their standing. This, for example, was a typical assessment of the shop trade union committee's role in distribution made by a worker from one of the in-mine transport shops: 'Our shop trade union committee doesn't do anything any more. It only distributes various things, televisions, fridges and so on. It gives them to some people and not to others. And all that comes of this is scandal and conflict.' Thus, the problems of the mine trade union committee are replicated at shop level: the shop trade union committees' responsibility for meeting workers' social needs does nothing to improve their standing.

It is notable that, when asked about the role of the shop trade union committees, neither trade unionists nor workers spontaneously mentioned representation or defence of workers' interests. The shop trade union committees are in fact in no position to represent the concerns of workers either to the mine trade union committee or to line management. Shop trade union presidents reported that they held shop trade union meetings on average four times a year – and even these were largely formal affairs. In general, the only contact they had with the workers in their shops, aside from their immediate work-mates, was when they were approached for some kind of material assistance. Shop trade union presidents are occasionally called upon to intervene in disputes, but this is far from a day-to-day occurrence.

Their most important duty as far as defence of workers is concerned is to give their consent or otherwise when management propose to dismiss a worker. This can be either an entirely formal process or one which involves negotiation: the trade union does not automatically take up the case of its member. Whether or not this occurs depends on the confidence of the shop trade union president and her relationship with the shop chief, as well as the character, reputation and situation of the worker in question. One miner summed up the attitude of his shop trade union committee to dismissals in the following way:

> In our shop there was a concrete case: one worker came to work drunk, and the girls [in the *lampovaya*] didn't let him go down the mine. So he ended up being marked down for absenteeism. The shop chief wanted to sack him. But the trade union stood up for him. He's got a family, children. They proposed that he should be transferred to lower-paid work and the shop chief had to agree. They didn't sack him . . .
>
> The trade union can stand up for a normal person. But if, for example, I haven't just skived off one day but several, and I work badly, then the trade union won't stand up for me. They also look at who they are going to speak up for and who they are not. Sometimes they might on the contrary tell management about a person who wags off work and works badly. And they will only support the dismissal of such a person. But before management sack or punish a person they have to consult with the trade union.

What this quotation illustrates is that in its attitude to dismissals, the union has changed little since the communist era. The position described by this miner is very similar to Mary McAulay's portrayal of the position of the unions in the past, which was discussed in the first chapter: the union will put the interests of the enterprise as a whole – its productivity and workplace discipline – before the interests of an individual worker. A shop trade union committee will not defend a worker whom it regards as shiftless against dismissal, because it is in the interests of the enterprise to get rid of such workers, but it will defend a 'normal' family man, because his family is part of the mining community and it is in no one's interest to foster avoidable desperation and anger in the mine's midst.

Thus, co-operation between the trade union and the administration continues down to shop level: their proximity to workers does not lead the shop trade union committees to take a less equivocal stance. Again, this is by no means a situation unique to Taldym. Nachalova, in the article cited above, claimed that in the reformed Rosugleprof shop trade union committees had a new role: to defend workers' interests. Nevertheless, her comments made it clear that she actually did not envisage the shop committees coming into conflict with shop management. Indeed, she did not seem to be sure whom workers should be defended *against*, since she presented work collectives as sites of communality where all essentially share the same goals. She therefore reduced the role of the shop trade union committee to that of an emollient, smoothing out incidental difficulties rather than systematically representing workers' interests. She argued:

> When we say that the trade union works well in this or that collective, we immediately imply something about the activities of the work group; [that it is a collective] where, as in small collectives, everyone knows everyone else not just as a good or a bad worker, but also from a moral and ethical point of view. Team work, interaction with others [*obshcheniye*], make a big contribution to the creation of a normal micro-climate in the collective. Is it normal everywhere? Far from it. From the outside perhaps [it looks as if] everything is going smoothly, but actually there's someone nursing a grievance against the head of the workshop, the shop chief and the trade union leader, because they didn't pay him attention at the right time, they didn't help him.

In this passage Nachalova links the shop trade union president and the line managers together as a bloc of authority figures and does not indicate any distinction between them. She moreover implies that the role of the trade union committees is to foster the right moral atmosphere in collectives –which, far from being about the defence of workers' interests, is very close to the disciplinary role that the shop union committees used to fulfil. Although, as explored in the last chapter, the tendency to identify the interests of workers with those of the enterprise does have an objective basis, the fact that a leading article in a union newspaper can advocate such a conception of the work of the shop union committees indicates that it is also a deeply ingrained perception which may well outlive the conditions which gave rise to it.

But it is not simply that the shop trade union presidents, informed by the same conception of the labour collective as the trade union president, run their committees along the same lines. The mine trade union committee, in its concern to avoid fostering conflict between workers and managers, does not actually promote the shop trade union committees as channels through which workers can express their grievances. To do so would be to place the trade union at the centre of conflict at the mine – a position which the previous chapter made clear the union is keen to avoid. One of the most striking ways in which this tendency manifests itself is in the union's preference for shift rather than shop trade union meetings. The union very rarely requests that the shop trade union presidents convene shop meetings, and consequently these are held on average only four times a year.

Meanwhile, the union only organises shift meetings when it has something particular to communicate to its members such as the timing or results of one of the national or regional demonstrations over late payment of subsidies. Workers are highly unlikely to raise any of the issues that concern them at shift meetings. First of all, such meetings are always rushed. They are convened before the beginning of the shift and the workers have to hurry off as soon as the main business is over. Miners especially have to make sure that they don't get to the *lampovaya* too late, so there is no space for stormy 'other business'. Most workers are, moreover, rather preoccupied with the shift ahead and want the meeting over and done with as soon as possible. Secondly, at shift meetings workers are simply thrown together with a random selection of workers from other shops. In such a forum they are unlikely to speak out, and will certainly not feel able to raise issues that only concern their shop. This was demonstrated quite clearly on a day of shift meetings in October 1994. A generally outspoken woman brigadier from the technical complex wanted to go to the meeting to raise the question of her collective's pay. Her work-mates encouraged her to attend, but finally she decided against it, saying, 'What am I going to do there – one *baba* [a colloquial term for an ordinary – originally peasant – woman] among all those men?' And her caution was fully justified: the only other woman at the meeting she would have attended (besides myself) was a member of the mine administration.

Shift meetings are thus easy to control. The workers confront the union and the administration as an unorganised mass, without any agreed agenda. The union and management, in stark contrast to this, always have a very clear idea about what they want to achieve by convening shift meetings. Moreover, shift meetings are specifically viewed by the union as a way of calming the workers, rather than mobilising them. The workers are given a formal chance to 'have their say' and let off steam, which means that it is harder for them then to complain that decisions have been made without consultation. At the same time, the meetings generally entail complex explanations of the mine's financial position, to which it is almost impossible for the workers, who are neither united nor prepared, to respond. The use of shift meetings as a means of reducing tension in the labour collective was demonstrated on the eve of the Miners' Day celebration in 1996. The trade union president was worried about what might happen over the weekend, because the mine could only afford to offer the miners 200,000 roubles as an 'advance' from their May pay packet.[2] (Surface-workers were to be offered 100,000.) He claimed, 'This weekend we are going to be sitting on a volcano. Any shift could strike.' He then continued, 'Probably I'll hold shift meetings tomorrow, to try and prevent it.' This is indeed what he did, and no strikes occurred over the weekend.

In this case the union did have a good reason to try to prevent a spontaneous strike: at the time of the meetings the co-ordinated national strike was still due to go ahead on the following Monday. Nevertheless, the incident highlighted the ability of the union to control shift meetings and to use them to contain conflict. As already noted in the previous chapter, the meetings were a perfect example of 'social partnership' in action: the trade union president and mine director put up a united front, firmly placing the blame for wage delays with the government as they explained why the mine was going to cease supply of coal (but not strike). At the same time, the mine president announced that, while he had managed to obtain money for miners and surface-workers to receive an advance, office workers and members of the administration would receive nothing. Workers seemed torn between scepticism and gratitude, but in spite of their doubts the meeting achieved its aim of preventing any kind of explosion.

Shop trade union meetings, in contrast to shift meetings, would give workers a far greater chance to get their views across. They could discuss the issues that concerned them within their collectives and ensure that these were raised at the shop meeting.[3] Then, at the meeting, it would be far easier for workers to request that the shop trade union president raised the questions concerned with the shop chief or the mine union committee. Workers within shops tend to know each other, and the shop thus provides a far more conducive environment for workers to express their grievances. The workers from the technical complex certainly felt this: whilst they were deliberating, on the occasion mentioned above, over whether to send their brigadier to a shift meeting, they bemoaned the failure of their shop trade union committee to convene shop meetings, which meant that their complaints were never heard.

In addition to inhibiting the interaction between workers and the shop trade union committees, the union's relationship with the mine administration also structures the relationship between the mine trade union committee and the shop trade union presidents. Rather than being concerned with matters of policy and strategy, the shop trade union presidents are treated rather like

[2] At the time this represented between a quarter and a tenth of their monthly wage (depending on the skill grade of the worker and the productivity of their shop).

[3] The trade union president does in theory recognise that giving workers the opportunity for greater involvement is the key to challenging what he portrays as their passivity. For example, when discussing the workers who had been involved in the collective agreement drafting and monitoring committee in 1995 he said, 'After working on this, they became completely different people; they began to understand a little bit. But I can't go round and talk to everyone; I can't get round them all.' By utilising his shop trade union committees, however, the trade union president would be able to 'get round them all': it is just that at present he does not see any need to do so.

dogsbodies responsible for the routine 'social work' of the shops: they are not required to be activists promoting worker involvement in union activity. Most of the trade union work at the mine is carried out by the seven-strong presidium of the trade union committee. The forty-odd shop trade union presidents (the number fluctuates somewhat) meet the trade union committee once a month, but the meeting is not so much a forum for consultation and exchange of information as a chance for the trade union president to keep the shop trade union presidents informed of his activities. Again, the president's desire to contain conflict within the labour collective is apparent in his addresses to these gatherings. He often takes the opportunity to explain a particular decision of the mine administration or to scotch rumours, and encourages the shop trade union presidents to disseminate the information within their shops. For example, at a meeting in August 1996 he devoted a considerable amount of time to countering the rumour (which is in permanent circulation at the mine) that members of the mine administration had been paid while the workers had received nothing for three months. The trade union president said that he monitored the payment of wages himself, had to sign for any payments which were made, and could personally guarantee that the administration had not been paid. He pleaded with the shop trade union presidents to make an effort to get this message across to workers, and to send any persistent doubters to him. He would then investigate their claims and ensure that anyone who *had* been paid in advance of the workers would be sacked. The intention behind this message, which he repeated even more forcibly at the shift meetings, is clear.[4]

Given the formal nature of the contact between the shop trade union presidents and the mine trade union committee, it is not surprising that many of the shop trade union presidents feel as remote from the charmed circle of the trade union committee as ordinary workers do. For example, one of the shop trade union presidents related a tale of injustice which revealed precisely that she felt as little 'part' of the union as most workers. As already mentioned, the housing waiting list is a focus of discontent at the mine, especially among the least privileged workers. The shop trade union president in question complained bitterly that after fifteen years on the housing waiting list she was still living in a one-roomed flat with her second husband, grown-up daughter and stepson, while in the same length of time on what is supposedly the same waiting list the trade

[4] The rumour persists in spite of the president's efforts. Whether or not it is true is difficult to prove. The mine director claimed that it was completely false, confiding to me, 'I'd be a fool to pay the administration and not the workers, it would be so easy for someone to find out.' This, however, is not the case: it would be possible to prove a formal payment, but there are all sorts of other ways in which managers could receive money from the mine, and these would be very hard for a worker to investigate. In any case, no amount of denial will convince the workers, as the comments of this woman worker illustrate: 'I read an interview in the newspaper the other day with a chief accountant of some enterprise. They asked her directly whether it was true that the administration got paid when the workers didn't. Well, of course, she said, no, we don't get paid either. She's got her privileges too – she doesn't want the whole world to know about them.'

union vice-president had improved his accommodation twice.[5] After relating this tale of injustice, this shop trade union president then began to complain in exactly the same way as most other workers about the inactivity and privilege of the trade union committee. This is in no sense to argue that the work of the shop trade union committees could be improved by extending the privileges enjoyed by the enterprise union committee to shop trade union presidents, but simply to illustrate the gulf between the union and its officers. As far as the shop trade union president was concerned, the mine trade union committee was nothing to do with her. She carried out social work: the mine union committee was, in her view, supposed to defend workers' interests, but instead was a bastion of privilege.

Conclusion

The relationship of the mine trade union with management thus defines the limits within which the shop trade union committees operate. They are not amenable to change 'from below' because they are not constituted as representative bodies: they are only ever used as a channel of top-down communication and even then not on a consistent or organised basis. They are not a channel through which workers can express their grievances; instead they function mainly as part of the social and welfare infrastructure of the enterprise. Thus, the increasing tension within Russian enterprises does not exert pressure on the shop trade union committees to change, because workers do not address their concerns to these bodies: in general they only turn to them when they require some form of financial or social assistance. Pressure from below will therefore only begin to influence the shop trade union committees if the mine trade union adopts a different policy towards these organisations and begins to use them as a two-way channel of communication with the workers. The constraints on the mine union have been discussed in the previous chapter: it is clear that its approach to the shop trade union committees will not change overnight. There is thus little prospect of a dynamic union reform movement emerging at the shop level in the near future. In the present environment, even were an explosion of worker activism to occur at the mine it would leave the shop trade union committees untouched on the side-lines.

[5] This particular injustice was something of a *cause célèbre* among the acquaintances of the shop trade union president: the comparison with the trade union vice-president was frequently invoked.

6

'Our mine': workers and the labour collective

The previous chapters have argued that the more active enterprise trade union committees, such as that at Taldym, are beginning to change. But the road to reform is a long one and the unions' progress down it, even if it continues, will not be rapid. Meanwhile, the independent trade union movement has atrophied: as already mentioned, the NPG now has very little influence in the Kuzbass region. Thus, at a time when not only their living standards but their whole way of life is under threat, workers lack effective collective representation. The next three chapters consider forms of collectivity and conflict within the enterprise, as well as workers' response to the privations imposed by reform. Does the lack of organised activity mask a history of subterranean struggle? Certainly conflict is increasing within enterprises, so how is it expressed? And why has increasing tension not prompted new attempts to mount a collective response to reform?

These questions will be dealt with in three stages. First, building on the analysis of the Soviet system of state paternalism developed in the first chapter, this chapter will begin to examine the characteristic forms of collectivity which existed within the Soviet enterprise. The chapter argues that the structure of the Soviet regime of production, which constitutes the labour collective as a supplicatory unity, and the internal dynamics of the enterprise to which the system of state paternalism gave rise, constituted a barrier to the definition and defence of workers' specific interests. Nevertheless, within the Soviet enterprise work groups were constituted as 'collectives', and the next chapter goes on to ask why it has not been possible for workers to realise the collectivist potential apparently immanent in this form of organisation. It takes one particular collective – the *lampovaya* – and examines the changing potential for collective organisation at the grass roots through an analysis of the history of the collective from the Gorbachev era onwards. The last chapter in the section will then ask: if workers are not organising within the workplace, or on a wider political level, how are they responding to reform? And how is their response influencing the development of the 'New Russia'?

Forms of collectivity within the Soviet and post-Soviet enterprise

Workers identify strongly with the ideal of the united labour collective despite their acute awareness of the deep division which exists within this supposedly solidaristic entity. This chapter examines this contradictory consciousness, and, on the basis of this analysis, begins to examine why workers' collectivist ideals are never realised in the form of independent organisation. The first section considers the form of workers' identification with the enterprise fostered by state paternalism. The second section then goes on to analyse the main forms of division within the enterprise and the form of oppositional consciousness to which this gives rise. The chapter argues that what above all characterises this consciousness is that it is negatively defined and is not expressed in any form of workers' self-organisation. The last section then attempts to provide an explanation for this limitation on the basis of the analysis developed in the first two sections. It also asks how far the structural barriers to workers' organisation identified by the chapter are being eroded in the transition period.

The labour collective

Soviet enterprise collectivism is a complex phenomenon. First, as already mentioned, the term 'collective' (*kollektiv*) is used to refer both to the whole workforce of an enterprise and to its individual subdivisions. Managers and workers use it to refer to the different work groups within the mine – the workers in the *lampovaya* would describe themselves as a 'collective', for example. The employees of a particular enterprise are also collectively referred to as the 'labour collective' (*trudovoi kollektiv*). In practice, the distinction between the two forms of collective in everyday speech is rather blurred, but in understanding the character of collectivism it is important to distinguish between them. To avoid confusion, hereafter the term 'labour collective' is used to refer to all the mine employees, whilst individual work groups are referred to as 'work collectives'. One further clarification should be made: 'the collective' is at the same time an ideological construct and a material reality and the term is used here in both its ideological and concrete senses. This reflects workers' usage of the term, which constantly shifts between the ideal and the real.

As argued in Chapter 1, the labour collective was the expression of the common interests of workers and managers in relation to the external administrative environment. This collectivity found very real expression in the authoritarian paternalist system of social and welfare provision, which is the tie that binds the individual worker and work groups to the labour collective. What interviews with workers at Taldym discussed within this chapter and the two following chapters reveal is that this reality was reflected in workers' subjective identification with the labour collective. The labour collective accordingly had a dual significance: it was at the same time the locus of social control and the locus of self-realisation, the point of intersection between the totalitarian aspirations of the Party-state and the individual and collective aspirations of the workers in whose name the Party-state ruled. This section explores the subjective significance of the labour collective during the communist period, before moving on to examine the form of enterprise collectivism which has evolved during the transition era.

In spite of its alienated form workers identified with and were attached to the labour collective of the communist era, which they saw as their immediate guarantee of security. This attachment was not unambivalent: workers were aware that the collective was riven by conflict, and they also resented the forms of control that the Party was able to exercise through the labour collective. For example, a persistent form of individual deviance which plagued the Soviet enterprise was drunkenness, usually on the part of male workers. The treatment of workers who were caught drunk (whether at work or, more usually, by the *militsiya* on the street) highlights the most repressive aspects of enterprise collectivism. If the level of inebriation was judged to be sufficiently serious then the offender would be detained and forced to spend the night in the drying out cell (*vytrezvitel'*). After this, punishment within the labour collective would be swift and severe. This is how one of the fitters from the in-mine transport shop described the process, in an impassioned impromptu speech to his work-mates, some of whom had been waxing lyrical about the communist past during a lunch-time political discussion:

> If you were a complete *alkash*, then they considered you to be ill, and the Party would look after you, send you to a clinic to get you treated. Ah, but if you were an ordinary, simple worker – perhaps you only drank once a year – but God help you if the *militsiya* caught you drunk: you'd have to spend the night in the *vytrezvitel'*, and then they shamed you in front of the whole collective [*pozorili na ves kollektiv*]. They'd remove you from the housing waiting list – you might be third on the list and they'd put you to the bottom – deprive you of a summer holiday, of your thirteenth month, your premium. You were completely shamed.

In such cases a notice would be put up at the enterprise naming the offender, informing the collective that he or she had had to spend the night in the *vytrezvitel'*, and then detailing all the benefits of which he or she had been deprived. This attempt to control the behaviour of workers (and male workers in particular) through withholding benefits from them and shaming them was deeply unpopular.

Nevertheless, despite such negative features, workers did tend to view the 'labour collective' as a valuable and meaningful entity: the very fact that being 'shamed in front of the whole collective' was seen as something that mattered is an indication of this. Indeed, although there was resentment about some of the forms of managerial control exercised in the name of the labour collective, other aspects of the system of paternalist provision which from the outside might appear repressive were actually appreciated by many workers and interpreted as a sign of concern rather than a form of control. This can be seen, for example, in the comments of the brigadier from the *zaryadnoe* about the system of social security delegations which existed in the past:

> Before when I was a member of the shop trade union committee we used to have social security delegations which visited people who were on sick leave. They went to them and gave them five roubles – it was possible to buy something with five roubles in those days. They were paid some attention and they got better more quickly because they felt that their colleagues were waiting for them at work, that they were needed. Now that doesn't happen. If you're off sick a friend from work might come and visit you, but that's all. In 1993, for example, I was in hospital for a month and no one came to see me except one friend from work . . .
>
> A person needs to know that he's needed in the collective; not just at home, but also in the collective.

This quotation neatly captures the duality of enterprise collectivism. Obviously, the social security delegations were not just designed to cheer the sick and suffering – they also functioned as a checking mechanism in the struggle against absenteeism. But this worker, who participated in the visits herself, interpreted them as a form of care. And in this she is by no means unique.[1] Such opinions highlight the importance to many workers of the sense of belonging that they derive from their membership of the labour collective. The collective is seen as a source of meaning and support in life: an individual finds her value within the collective. Being cut loose or neglected is bad for the individual and bad for society.

In the transition era this attachment to the collective remains strong. Many of the features of the past system of state paternalism, such as the social security delegations, have disintegrated. But social provision is still important to workers, and, as illustrated in Chapter 2, workers still turn to the enterprise in times of need. Moreover, the common interest of workers and managers persists. While in the past this took the form of the informal 'production pact' described in Chapter 1, now the labour collective forms a supplicatory unity in the face of the Coal Department of the Ministry of Fuel and Power (formerly Rosugol') and the government. This sense of the labour collective as an embattled entity is only strengthened by the threat to the coal industry's future. The attachment to 'our mine' is strong and in many instances can eclipse conflict within the labour collective. The fact that the workers are now the formal

[1] Such opinions were, however, particularly prevalent among women, who, to a certain extent, appreciated the attempts of the authorities to control the behaviour of male workers. For example, many women at the mine were actually quite appreciative of the mine Party committee's efforts to control drunkenness, especially in the days of Gorbachev's war against alcohol. As one worker recounted with approval, 'in 1985 there was a "war against alcohol". It was very strict. We in the Party sent a lot of people for treatment. Wives of men with drinking problems would come to the Party and ask for their husbands to receive treatment. . . . The Party disciplined people. This meant that people tried to show their best side.'

owners of the mine has only strengthened the case for seeing the labour collective as a united (and besieged) entity. This, for example, was a typical attempt of a worker to reconcile the existence of conflict with the ideal of the united labour collective. She resolved the dilemma in favour of the collective: 'Some people are threatening to take the mine to court for non-payment of wages. I don't think that's right. Now we're an AO – we are the shareholders. It works out that we're taking ourselves to court. It's hard to sort out, but I think that the problem lies with the concern and Rosugol'.' Another worker struggled with the same contradiction, although she did not come to a firm conclusion. She condemned the position of the trade union, and explained: 'The trade union should defend the workers and struggle against the administration, but it can't. Why? Because the director is elected by the *narod* (people) and the trade union can't go against the *narod.* It can't do anything. It just does the work of the administration.' In this case the idea of the united collective, personified in the director, is seen as a barrier to the expression of real conflicts, but the worker sees no way out of this: the trade union, in her opinion, can do nothing.[2]

The director is both the embodiment of this contradiction – he is elected by the workers, but is also the power against whom they have to be defended – and the means through which workers resolve it: they need to be defended against the director only because he is not fulfilling his role as the father of the collective. That is, 'one enterprise' collectivism is structured through the relationship to the paternalist director. Workers see as the answer to their problems not consistent collective pressure in all-mine fora, but directoral repentance or re-election. Thus, as the earlier discussion of privatisation indicated, though workers have theoretically gained control of the mine through privatisation to the labour collective, they have so far proved unable to realise this potential power in any other form than the periodic re-election of the mine director and his team of managers. The director is the figurehead of the collective and is supposed to look after it. Such feelings become particularly acute in relation to the 'outside world'. The director is expected to do battle against Moscow and other assorted authorities in defence of the enterprise. The mere mention of such struggles is enough to provoke workers to defend 'their' enterprise passionately. Meanwhile, the behaviour of both workers and managers is informed by this paternalist conception of the enterprise. At Taldym mine, for example, every Monday morning the director has a 'surgery' to which workers take their problems. People raise a variety of concerns, many of them connected with home rather than work. One worker, for example, needed some topsoil for her garden and hoped that the director could arrange for the mine to deliver some. Other people go and beg for their children to be taken on by the mine, which officially is not taking on any more staff. This is all seen as quite normal – being a member of the labour collective is about being looked after.

[2] This analysis is actually quite close to that of the last two chapters: since the trade union represents the whole labour collective rather than workers in particular, it is ineffective as a trade union.

Such a conception of the enterprise has its problems. Workers do not like to be made to feel their supplicatory status, but, so long as the collective retains its alienated form, workers are ultimately dependent for their guarantees on the will and the whim of their director. This generates an enormous amount of resentment. Two female construction workers claimed, for example, that the director could be curt to those who asked for the mine to pay the funeral expenses of relatives killed in industrial accidents. Such disappointments in the director are the cause of bitter denunciations. Workers continually feel that 'bad father' directors are failing to fulfil their obligations, especially since their concept of paternalism is significantly wider than that of the management. For example, a typical complaint from a woman worker concerned the fact that the mine management had refused to help a miner whose house had burned down. He had two children and his wife was pregnant. She thought that in this situation it was the duty of the mine to spare a brigade of workers to rebuild the miner's house before the baby was born. Indeed, since the director offered only a pittance in compensation, the workers were forced to stand in for the inadequate paternalist and they organised a collection of money and goods for the family themselves. Such tension results from the fact that workers and managers have differing concepts of collectivism. The workers, rather than thinking of themselves as dependent, feel that management has a duty to meet the needs of the labour collective in accordance with egalitarian principles of social justice. Managers, meanwhile, have an authoritarian relationship to the collective and believe that it is their prerogative to decide the level and distribution of provision within the enterprise.

Although the paternalist enterprise is being eroded by reform, the labour collective is still a site of social provision as well as a social focus and source of meaning in Vishnovka. In the mind of the workers, the collective should be run on principles of egalitarian collectivism, administered by a just director. The director, on the other hand, views the enterprise as *his* enterprise, which he rules as an authoritarian paternalist. This gives rise to the perennial contrast between the workers' ideal vision of the enterprise and the everyday reality, a contrast which the workers attribute not to the paternalistic system but to the failings of the director. The solution, in the eyes of the workers, is not to free themselves from their dependence, but to change the director. The workers' aspirations, therefore, are never channelled into collective organisation, but are always focused on the person of the director.

Blat

While workers' ideal vision is that of the united labour collective, their daily experience reveals the enterprise to be deeply divided. This section examines the form and basis of divisions generated within the paternalist enterprise, focusing on questions of distribution, which emerged from the interviews as the issue which caused most frustration among workers. For, as argued in the first chapter, what above all characterised social provision within the Soviet enterprise was its inadequacy. Within an environment of chronic shortage and competition for resources, money, as has been seen, played only a marginal role in regulating access to goods and services. In its place informal mechanisms developed to determine distribution, the most significant of which was *blat*. This feature of the life of the Soviet enterprise was a major source of dispute and division within the supposedly united labour collective. This section will explain the working of this system, which still orders the life of the enterprise. This is followed by sections highlighting the relationship of *blat* to other forms of division and analysing how far such systems are being transmuted by transition.

Blat is the Russian term that is used to describe a particular form of informal relations through which people create and call in obligations in order to secure access to a variety of desirable objects, from jobs to caviar. People use the means at their disposal to do a favour for someone in anticipation that this person's services may be useful in the future. One woman worker at Taldym mine gave the following example to define *blat*: if the mine trade union president goes to the local shop to buy caviar and the shop assistant has old caviar and new caviar, she will make sure that the trade union president gets the best stuff. She will do this because the trade union president controls, among other things, the housing waiting list and she may need a favour from him in the future. Since there are no precise 'rates', *blat* is a somewhat uncertain currency, but the implied obligation of the generous gesture is always understood. For example, the former mine director, criticised by an opponent at a meeting of shareholders' for spending the mine's money on building garages for the local police, responded by saying, 'any director would be a fool not to do this'. No further explanation was required – all the workers at the meeting understood that the good will of the police was necessary, not only to allow for any personal peccadilloes of the director, but also for the smooth running of the mine. *Blat* also includes pre-existing relations of kith and kin, which in the Soviet setting imply automatic obligations: anyone in a position of influence in an enterprise is expected, as far as possible, to provide work for close friends and family; shop chiefs within enterprises are under an informal obligation not to blame any one close to them for any mistakes they make at work; shop workers are expected to obtain any desirable goods they happen to be selling for family and close friends, and so on. Thus, anyone with something to offer or with the right connections can become part of the *blat* system. Most people can participate in the system to at least some extent through their family and friends, but *blat* is in general a hierarchical system in which position and proximity to desirable resources are all.

The degree of *blat* open to different individuals thus varies greatly, and in this way it divides people into concentric patterns of insiders and outsiders, or what in Russian are referred to as *blatnye* (those to whom *blat* is available) and *neblatnye* (those to whom it is not). Those who have any position of power within the enterprise, such as members of management or the trade union president, automatically become *blatnye*, independently of their will. *Blat* is a property of their position rather than their person. When they lose the position, they, like everyone else, have to depend on their family and friends, though they are of course more likely to have friends in high places. In this sense, the boundaries between the *blatnye* and *neblatnye* are fluid: a good position can be lost, and an insider in one situation can be an outsider in another.

Nevertheless, to the ordinary workers, who in most circumstances fall within the ranks of the *neblatnye*, the system seems rigid and impenetrable: they feel themselves to be forever on the outside. They feel disempowered and diminished by *blat* within the enterprise; a feeling that was powerfully described by one worker who said that it was only the *blatnye* who were 'treated like human beings'. This sentiment was very common: a miner, talking about the way he felt the trade union committee used the mine's leisure facilities for their own private entertainment, described the atmosphere at such gatherings using the potent phrase, 'Everything's done in a human way [*po chelovech'i*] there.' He then continued 'and for the miners . . .' before breaking off in disgust, but the implicit contrast was clear – the *blatnye* create a civilised world for themselves while the miners are not even treated like human beings.[3] Such resentment pervades the enterprise like a cancer: it is difficult to talk to a 'simple worker' at Taldym mine without some reference being made to the frustration caused by the unfairness of *blat*.

During the communist era paternalist provision within the enterprise was the focus of this frustration: workers deeply resented the discretionary basis on which benefits were distributed. Although this, as will be discussed below, is beginning to change, in the past workers' quality of life was chiefly determined by their position in the various queues at the enterprise. And those with connections always came first, as this worker from the *lampovaya*, who is in her mid-fifties, explained:

> I, for example, may have been first in the queue, but you're *blatnye* so you got your flat first. Or holiday vouchers – perhaps I went to the trade union and asked for them before you but you got the holiday all the same if you had *blat*. And kindergarten places, that was another hot issue in the past. . . . Those were the three main questions in the past: the distribution of flats, holiday vouchers and vouchers for the *profilaktorii*, and the waiting list for the kindergarten. And now the main question is who gets paid on time.
>
> Oh, and the waiting list to buy a car – that was another place where *blat* was everything. Up until about 1993 it was very difficult to buy a car. Now if you've got the money it's no problem. But before you needed to wait for years. The only people who got them were those close to the 'distribution kitchen' . . .

[3] Similar examples included the idea that managers and office workers – sometimes referred to as 'white' people – would always be given the food of their choice in the mine canteen, whereas as soon as a dirty worker – a 'black' person – came along she would be told '*nechego netu*': 'there's nothing left'.

> You could [also] be removed from the housing waiting list or the waiting list for the kindergarten, for example, if you violated discipline. But again, it depended on *blat*. If one of the *blatnye* violated discipline, he wouldn't get removed from any list. The same is true now. In short, if you've got *blat*, you don't have any problems.

In the communist era, the sense of humiliation felt by 'simple workers' was not confined to the enterprise, it was also felt outside work. The *neblatnye* felt that they encountered injustice at every step. This, for example, was the comment of one male mineworker on his experience at school:

> When I was at school no one helped me. If I was sitting there unable to do my sums, the teacher would leave me to it and I would end up with a *dvoika* [2]. But the mine director's son – the teacher would go and help him and he'd get a *pyaterka* [5]. [Five is the highest mark in Russian schools and one is the lowest. A two is considered pretty shameful.] Does this get to me? Yes it gets to me all right.[4]

Such feelings are commonplace. The problem was obviously particularly acute in small settlements such as Vishnovka, but it was not only in such places that the subordination of workers within the enterprise carried over into other spheres of life. Enterprises everywhere played a central role in service provision; even in large towns workers tended to be clustered in the housing estates around their enterprises. And even when such urban enterprises were part of a wider municipal system, they none the less formed something of an enclosed world.

Although in the present period elements of the state paternalist system, such as the housing waiting list, are losing their salience, position and connections are still very important at the mine. For example, now that the Communist Party is no longer there to exercise control over managers, workers have a strong sense that *blatnye* managers at the mine are using their position to enrich themselves at the expense of the workers. This, for example, was the response of a group of miners to the question of who they defined as *blatnye*:

> MINER 1: It's the mine administration. There are lots of vice-presidents, it's all these 'vices' [vice-presidents]. They are all *blatnye*. They can get all sorts of resources; they can build themselves houses. And the chief accountant: she pays us a bit less and then keeps it herself. She can rob the miners.
>
> MINER 2: Yes, the chief accountant gives the miners' money to her friends.

Such feelings of being cheated by management are very common, as is the vilification of female office workers, who are often portrayed as dishonest embezzlers by manual workers.[5]

[4] Again, a version of the powerful rhetorical expression '*Mne obidno? Obidno*', was used here.

[5] This contrast between the deceived but morally virtuous 'simple workers' and parasitic, corrupt administrators is part of a long tradition in Russia. Steve Smith notes that the expression '*nas obmanuli*' – they deceived us – was a *leitmotif* of workers' protests in 1917 (Smith, 1997: 5), while Davies argues that workers in the Stalinist era based their claim to moral superiority on the

Moreover, *blat* still plays an important role in the conduct in ordinary life. This is neatly illustrated by one worker's account of her attempt to rent a tractor from the mine in the summer of 1996 to harvest her hay. The daily rent for tractors at this time was 400,000 roubles: approximately half this worker's monthly wage.[6] She rents a tractor every year, but usually she arranges this through her connections: although she, like everyone else, laments the injustice of *blat*, she is actually very well connected at the mine. In 1996 she decided to do it 'like a simple person' and see how it turned out:

> Every day I turned up to ask and they'd say 'you'll have it tomorrow'. It's offensive [*obidno*] – I've paid half my wages for it. Anyway, finally on Friday I went to see the chief mechanic . . . and he said 'Why on earth didn't you come before? You'll have it on Monday'. I said, 'Well I just thought I'd try and do it like a simple worker.'

The moral of this is clear: doing anything 'like a simple worker' is time-consuming, irritating and sometimes fruitless.

Moreover, workers claim that *blat* has extended into new, highly significant territory: they argue that it now determines the order in which workers receive their wages. In an era of endemic wage delays workers claim that the well connected are given priority whenever there is any hard cash (*zhivye den'gi*) at the mine. As mentioned in the previous chapter, it is very difficult to prove whether or not this is the case, but workers are convinced that it occurs. The following accusation from a worker from the *lampovaya* is typical:

> Of course, when we're paid late, the *blatnye* can still get money. When some money arrives at the mine, their contact will tell them that there's money at the *kassa* [cash office] and they'll go and get paid. . . . A month's pay, half a month's pay, depending on how much money there is. Say, for example, a worker from the *lampovaya* has a friend in the accounts department – she'll phone her and say, 'Go to the cashier now and you can get some of your pay.' Well, of course, this creates a lot of bad feeling – that some people have been paid and others haven't. The ones who haven't start to blame the ones who have for their problems. And the ones who haven't [been paid] are in the majority. This is making people more and more angry as time goes on. The problem is getting more and more acute.

This is certainly a major cause of tension at the mine: the efforts of the trade union president to scotch such rumours, described in the last chapter, reveal that it is viewed as a potentially explosive issue. The delicacy of the question is highlighted in the following account of the ill-fated outburst of one worker about this perceived injustice:

assumption 'that toil in itself is redemptive and suffering is good. Since those in power did not actually work, they became morally corrupt' (Davies, 1997: 82). Given that workers privilege manual over mental labour – another long tradition deriving from the peasant past – office workers are generally included in the ranks of the idle parasites in workers' mental categories.

[6] The money is deducted from workers' wages: they are not expected to pay cash for such services.

> If you've got *blat* then you can live, but if you haven't then you just survive. That's what they say here and it's true. For example, the management – we've noticed that they don't queue at the bank to get their pay with us. They get it directly from the *kassa* and no doubt about a month before us. One of my friends from the technical complex, she said that to the director's face – she said, 'it's OK for you, you don't get paid late'. He didn't like it. What she said was true – but one thing I've learned is that people don't like it when you tell them the truth to their faces. Now she says they are after her. They check up on her every movement – whether she's late for work or leaves early – he wants to catch her out. He wants to punish her. Supposedly we've got democracy now, but it means nothing.

Thus, the focus of discontent at the mine is shifting somewhat, but the idea that the labour collective is divided into a 'them and us' of 'simple workers' and the *blatnye* remains as strong as ever.

The privileges of Party members

Two distinctions should be introduced into this picture. First, with reference to the past, it is important to explain how Party membership fitted into this system of personal ties. Then, in the next section, the evolving relationship between *blat*, bribery and money will be discussed.

The meaning of Party membership in the Soviet context can be elucidated by comparing it with the situation which existed in China. Andrew Walder has analysed the system of factory administration in communist China, and concluded that a key element in the prevention of collective action was the deliberately fostered division between the 15 and 20 per cent who were Party members and ordinary workers: 'the party itself rewards and promotes people preferentially according to the loyalty and service they render to management and party', a form of clientelism which Walder terms 'principled particularism' (Walder, 1986: 25). Although personal loyalties do arise between specific leaders and followers in this setting, this personal dimension is not the significant feature of these ties: 'Party-clientelism is created "from above"; it is an institutionally prescribed clientelist network' (25). Alongside this system there is a network of instrumental-personal ties through which primarily non-activist workers pursue their interests. These relationships, popularly referred to by the term *guanxi* in China, are comparable to those referred to by the term *blat* in Russian (171). These form an important component of the system of factory administration just as they do in Russia, but the particular feature of the Chinese industrial system which distinguishes it from that of other communist countries is the strong system of party-clientelism.

In the Soviet Union, Party members did, as in China, have privileges, but the Party did not control the enterprise in the same way as it did in China, and neither was Party membership accorded the same ideological significance.

Walder's study shows that the Chinese communist authorities had a far greater concern for political mobilisation and intensive group organisation than did their Soviet bloc counterparts. Although this occurred to a degree in the Soviet Union of the Stalinist era it was not continued thereafter.[7] The form of Stalinist industrialisation dictated a less politicised approach to factory administration. First, labour turnover rates in Russia were very high, while in China the work force was highly stable, which allowed deeper Party organisation. Second, while in the Soviet context factory Party organisations had to be rapidly created when new enterprises were established (as occurred at Taldym), in China the party organisations were already stable and well developed in the era of rapid industrialisation: in contrast to the 'disciplined elite' created by Lenin, the Chinese Communist Party was 4.5 million strong when it achieved power. Moreover, the turnover in the enterprise Party organisations in the Stalinist era was nearly as high as among workers, because the weakness of the Party in rural areas meant that newly recruited members from industrial enterprises were often almost immediately dispatched to help with the collectivisation campaign. Thus, in the Soviet context, rather than the Party being responsible for discipline, managers controlled workers through their command over the distribution of goods and services and their ability to manipulate wage differentials and piece rates. In China, the Party was the key institution in the enterprise, and industry was run on the basis of mobilisation through the Party (Walder, 1986: 113–20).[8] In contrast to this, Soviet managers, even though they were nearly always Party members, acted in partnership with, or occasionally in opposition to, the enterprise Party secretary.[9]

In the Soviet enterprise, Party members, unlike their Chinese counterparts, did not have to demonstrate high levels of political consciousness, and they

[7] Soviet mobilisation campaigns were always more instrumental: indeed, political campaigns which interfered with production were generally prohibited. In China, by contrast, the Party was utterly serious in its attempts to re-educate and resocialise workers. Thus, for example, while 'group criticism' was a pervasive element in Chinese accounts of organisational life of the 1950s and 1960s, it did not feature nearly so prominently in accounts of the Stalinist period. Criticism and self-criticism meetings appear to have been held only among Party members and even then not very often (Walder, 1986: 121). The Chinese authorities, in contrast to their Soviet counterparts, 'infused factory reward systems with a pervasive moral-political content': 'to the student of China the Soviet experience reflects scant attention to the moral and ideological cultivation of workers' (121–2). Stalinist campaigns directed at workers focused on dedication to work: the Party was not interested in workers' inner beliefs regarding the latest ideological twist, so long as they showed no sign of open resistance. For an account of the work-focused campaigns of the period see Kotkin, 1994.

[8] The Chinese initially attempted to introduce Soviet-style incentive systems in the early 1950s, but they were administratively complex and required large numbers of educated clerical staff, who were in short supply. Such problems could have been overcome, but it was not actually in the interests of the factory Party organisations to follow the Soviet route, which would have reduced their power in relation to that of the technicians.

[9] The director was, however, unquestionably subordinate to higher Party organs.

were not set aside from other workers to the same extent.[10] Instead, Party membership was a way of co-opting energetic, respected or charismatic members of the workforce and ensuring that their leadership qualities were exercised in favour of the authorities.[11] Moreover, what was primarily valued in Party members was not their political commitment, but their work record: the Party was above all concerned to promote production and its members were supposed to set a good example in this regard.[12] Thus, the Party sought to recruit good workers: it was considered important for the Party's authority that they were members. This can be seen in the remarks of an MGVM from Taldym who was active in the trade union and Komsomol. He had a glowing record of achievement, and the Party was keen to enrol him:

> Communists had big privileges. . . . I worked and to this day I haven't become a communist. And I worked well, and the whole time they said to me 'Vyacheslav Ivanovich, don't you want to join the Party? You work well. Why aren't you a communist?' I always gave the same answer, 'I want to show that I'm not a communist and I don't work any worse than a communist.'

Often the Party attempted to lure such workers with the promise of promotion or other benefits. A skilled and responsible male worker from one of the mine maintenance shops reported:

> If you joined you got promotion. I was on skill grade four, and the shop chief openly said that I could move up to five if I joined the Party. But I refused; it was a principled decision. The communists shot my grandfather. He was what they called a kulak. I stayed on skill grade four for years.

Acquiring a flat was the biggest inducement the Party could offer: Party members were given preferential treatment on the housing waiting list and help if they wanted to build their own home in the private sector. As one worker from the *lampovaya* explained:

> Party members got flats first; they got any material they needed to build a house first. They got everything first. . . . The *blatnye* were those who had power, and their friends and relations. But the communists had privileges. . . . For example, a friend of mine was on the housing waiting list and they just weren't giving her a flat. She'd been on it for ages. So eventually she complained to the Party committee. They said, write an application, join the Party and you'll get a flat. So she did, and they gave her a flat. Then, two weeks after she'd got all the documents, she just left. Good on her.

[10] In China activists were expected to denounce their fellow workers and were hence deeply resented. They were also isolated, which reinforced their dependence on the Party. In contrast to this, denunciations of co-workers by Party members seems to have been the exception rather than the norm in post-Stalinist Russia.

[11] After the 1989 miners' strike there was a concerted effort to recruit 'active' workers into the Party.

[12] A content analysis of the minutes of Party bureaux and committees in ten enterprises in the Komi Republic between 1966 and 1975 found that on average questions of production management occupied 62.4 per cent of the meetings (Ilyin and Ilyina, 1996: 375).

Despite these privileges, however, Party membership did not divide the workforce in the dramatic way that it did in China: in the Soviet context, the networks of instrumental-personal ties, which everyone had to cultivate to at least some degree, were far more divisive.[13] Although, of course, becoming a Party member could be a means to develop such ties, this is distinct from the system of 'principled particularism' depicted by Walder. A retired (female) ITR – a former production engineer – described the relationship between *blat* and Party membership in the following terms:

> Being a member of the Party created *blat.* People went into the Party in order to obtain privileges. But there were people outside the Party who had *blat.* Managers were *forced* to join the Party – for example, all the shop chiefs had to join. Take shop number eight: all the *blatnye* worked there because it was the best shop in the mine. I can stay close to home with my example: all my sons worked in shop number eight because I was the production engineer. The Party gave me a lot.

Thus, position, Party membership and *blat* were mutually reinforcing elements in the enterprise hierarchy, although *blat* could exist independently, as, for example, in the case of those connected to managers or Party members. The fact that Party members had privileges was not hidden and was accepted as an, albeit undesirable, fact of life: at least the Party members had to pay for their privileges by attending meetings, presenting themselves as model workers and so on.[14] *Blat*, by contrast, was an open secret which drove workers nearly to distraction.

[13] Alongside the hierarchies of access, there were also status hierarchies between different categories of worker within enterprises. These hierarchies were related but did not fully coincide. For example, in the era of state planning (though less so now) the status of those main production workers who 'drove on' the plan was higher than that of auxiliary workers who were generally paid less, while the elite of the enterprise were the so-called *kadrovye* workers who had a long record of service, high skills, a good disciplinary record and some record of voluntary 'social' activity (Kozina and Borisov, 1996: 136–61.) Nevertheless, even though, for example, workers in the elite production collectives could expect certain privileges, divisions between main production and auxiliary workers did not fully determine their respective levels of welfare: at Taldym there are well-connected women workers in auxiliary collectives who have much better access to resources at the mine than many of the face and development workers who constitute the elite workers at the mine. This non-correspondence between different hierarchies may be more common at an enterprise such as Taldym, which is situated in a small settlement where blood ties or (culturally crucial) godparent relations between workers and managers are more likely to exist, but it is by no means confined to such enterprises.

[14] Moreover, Party membership was quite accessible to those who had a decent work record. Indeed, at Taldym in the late communist period the Party seemed to be keen to have anyone who was prepared to join. I spoke to several 'good workers' who refused to join, although not enough to provide a definite opinion of the reasons why workers were reluctant to join the Party. On the basis of impressionistic evidence, however, it seems that workers did not want to be burdened by the extra duties that Party membership involved. Moreover, although Party members mixed with other workers quite freely within their collectives and were not stigmatised unless they made themselves objectionable in other ways, many workers none the less felt that to join the Party was to sully their status as 'simple workers'. Their view of the Party was contradictory, however. On the one hand, as

Bribery

Blat should be distinguished from bribe giving. Bribes are resorted to where people have no recourse to *blat*. The worker who, for example, desperately needs to receive her sick pay or holiday pay on time (payments which at Taldym mine seem to be made on the discretion of the employees in the accounts department), and has no connections and no influence which would induce the accounts department to do her the 'favour' of paying her on time, would have to resort to a bribe in order to be paid. Bribe giving is, however, a less universal and refined system than *blat*. Everyone is aware of *blat* and knows the rules of the game. Interestingly, though it is generally recognised that bribing is a possibility, not everyone is aware of the etiquette. For example, two workers from the *lampovaya* at Taldym, who knew and liked each other, had very different attitudes to bribe giving. One of them saw it as a humiliating, but nevertheless necessary, activity, and reported that she had bribed the accounts department the week before in order to receive her husband's sick pay, which the accountant had 'suddenly' found that she was able to pay when she received a packet of high-quality imported biscuits. The other worker said that she had no idea how to go about giving a bribe:

> I just don't know how I could walk into an office and give one of those women a bottle of champagne so she'd help me. How would I do it and what would I say? I just don't know. And how could she say, 'right, yes, I can help you', when the day before she'd sworn blind there was nothing she could do? It would be so blatant. I just couldn't do it. I would be ashamed.

This difference of opinion on bribery did not extend to the subject of *blat*: both women felt equally excluded by that.

Having said this, however, there is some evidence that impersonal forms of exchange are becoming more important as economic reform progresses. Although the Russian economy is not yet fully monetised, money is beginning to play a more important role in the system. A whole range of goods and services, which could previously only be acquired through the enterprise or connections in the distribution network, can now simply be bought for money: flats, cars, holidays and erstwhile 'luxuries' such as condensed milk. Of course, workers who receive their unindexed wages several months late do not have much chance of buying a flat on the open market,[15] while acquiring a car is a

the approval of the worker who cheated the Party to get a flat indicates, it was viewed as part of the hostile 'them'. But at the same time it was viewed as a weapon and a protection against 'them' in the mine administration: the worker who applauded her friend's trickery also spoke with approval about the way that the town Party committee could 'punish, and punish hard' members of the mine administration who were caught making mistakes. This parallels workers' attitudes to managers, who are often portrayed as their defenders in relation to the 'outside world', but their oppressors within the enterprise.

[15] To buy a flat *pod zarplatu* (through deductions from wages) through the enterprise it is still necessary to be at the top of the housing waiting list.

major struggle. Moreover, the fact that the payment of wages up to six months late does not lead to mass starvation and riots shows the extent to which money is still not what determines the ability to survive in Russia. None the less, it can now open a whole variety of doors to which *blat* previously constituted the key. This has made many small details of life, in particular visits to food stores, far easier, but in general workers feel that one form of exclusion and injustice has simply been exchanged for another, with the winners and losers remaining largely the same. The comment of this male worker neatly captures this feeling:

> It's very hard. My daughter tried to get into college here. . . . They said to them openly, you're all grade threes and so if you want to get in you're going to have to pay – well, she'd been getting fours and fives at school, but it makes no difference. Before you needed *blat*, but now money determines everything. You have to pay. So the mine administration will all make sure that their children are properly educated just like they were before: the mine can pay. . . . The only thing that has improved is that before the shops were empty and you needed *blat* to buy anything. Now you just need money.

Others felt that money was gradually replacing the uncertain currency of *blat*, and that bribery was now more pervasive than *blat* had been in the past:

> It's even more unjust now than it used to be. Now bribes are becoming more important than they were in the past. It's impossible to live honestly these days. '*Chestnyi*' [honest] is a word which will soon disappear from our language . . .
>
> Under communism there were boundaries. You knew that you could go thus far and no further. . . . *Blat* always existed of course, but there was some level of control. You could complain to the Party committee. Now there is no control at all.

Thus, although the forms of division may be changing within the enterprise, workers feel as excluded as they ever did.

The limits of collectivism

From this standpoint the ideal of the harmonious collective is shown up to be a sham and a deception: the enterprise is clearly divided into 'them and us'. And, broadly, this 'them and us' means managers and workers. Thus, though workers at some level aspire, on behalf of their enterprises, to the ideal of the united collective, they also have a strongly developed oppositional consciousness.[16] Indeed, their critique of their society could be described as 'total' in

[16] Burawoy and Lukács (1992) have argued, with reference to Hungary, that the regime of production under state socialism was more conducive to the development of class consciousness than that of capitalism, claiming that 'class consciousness is endemic to socialist production' (83). They argue, however, that the consciousness produced is 'negative', in that it is purely oppositional, and that a 'positive' consciousness, 'a vision of an alternative order . . . can only be forged in class *mobilisation*' (114). It is this step from 'negative' to 'positive' class consciousness which proves so elusive in the post-Soviet context.

that it is not confined to the workplace: workers see the world in terms of a hierarchy in and outside work in which they are definitely at the bottom. Notwithstanding the misogynist undertones of the animus against female office workers, it could be argued, then, that *blat* engenders a consciousness in workers which is radical and even potentially revolutionary.[17] But there is a major obstacle on the road to revolution: the community of the *neblatnye* is only defined negatively in relation to the enterprise 'insiders', and as such has only an emotional and rhetorical existence.

The apparent unity of the 'simple workers' in opposition to their *blatnye* managers conceals other foci of collective identification, as well as divisions between workers. The definition of a 'simple worker' can be a slippery issue and hence the boundaries of the 'them and us' are constantly shifting. As with *blat*, an individual can be an insider in one situation and an outsider in another. And although in general workers assume a unity between themselves and colleagues in non-managerial, dirty, manual, unpleasant or unskilled jobs, such a collectivity can shatter in concrete situations. The following extract, from a face-worker at Taldym mine talking about the decision of a trade union meeting to stop paying its female employees for a monthly 'women's day' at home, gives an interesting insight into this process:

> Up until then [the last trade union meeting] the collective agreement gave women one paid free day. At the expense of the mine, at my expense, at the expense of my comrade. It's bad is it, that my wife has one free day to spend with the children? I'd give her half my wages. Let her spend time with the children. That's not how it worked out. One clever clogs got up – perhaps he didn't have children – [and said] 'why should I give my money to that lot?' . . .
>
> How many women are there at the mine? You can count them on one hand. The office workers, for instance, well, what do they do? And then there are those [collectives] such as the *lampovaya*, the technical complex and the *zaryadnoe* where the conditions are diabolical . . .
>
> There are normal men . . . and then there are men who for a kopek are prepared to take money away from a comrade at work. . . . They think like this, 'they sit there in the *lampovaya* and they don't do a blind bit of work. They sit there in the *lampovaya*, I'm fucked if I'm going to give them my money.' Yes, they sit there in the dust and the dirt . . .
>
> Of course men, they think about women at the mine: the office workers sit there, three storeys of them and they fleece us. . . . Someone should have stood up and said 'Don't pay the office workers'. But there are women who have labour-intensive work, dirty [work]. They should have been paid.

[17] Workers' denunciations of injustice would certainly excite anyone looking for evidence of revolutionary potential in the Russian working class. One worker, for example, after denouncing the division of the mine into 'black' and 'white' people, was explicit in her intentions: 'The *chinovniki* [officials] are bourgeois. We need to get them all against the wall and shoot the lot of them – and I want to watch'. David Mandel's published interviews also contain many extracts which highlight the rhetorical radicalism of Russian workers and union leaders (Mandel, 1993).

This extract gives a good indication of the shifting nature of the workers' collective identification. Some male workers, it is argued, do not see women in collectives such as the *lampovaya* as real workers and assume the women sit around all day. Since the women lost their paid 'women's day' at this meeting it can be assumed that some male workers do hold this view. Other workers, such as the speaker, have a different view. Women who do dirty work in diabolical conditions are worthy of respect. All are apparently clear on one thing: the office workers are not simple workers, they are parasites. Although definitions of the worker camp differ, this potent division between simple workers and the mine administration is felt by all. Nevertheless, amid the confusion over collective identification, women workers lost out. Their defeat illustrates the limitations imposed by the fact that the seemingly passionate solidarity of the *neblatnye* is only negatively defined. Although continually reasserted verbally, the collectivity of the *neblatnye* dissolves in the face of other cleavages such as that between miners and female surface workers illustrated above.

Such divisions are not even confined to different categories of worker. For example, female surface-workers from different collectives display a striking lack of solidarity with each other. This was dramatically illustrated at the 1995 meeting to review and revise the collective agreement at Taldym. First of all, the trade union president proposed that the lowest-paid workers at the mine – the women who work at the *profilaktorii* – should be the first to be paid the back pay that they were owed because, he claimed, it was impossible to survive on such low wages if they were paid late. This proposal was agreed to by the meeting, but not before a few derisory comments had been made by other women workers: 'We could all get up and tell the meeting our monthly wage and argue that paying such low wages late was impossible.' Then, during a different discussion on pay, a representative from the *kotel'naya* stood up and complained about the pay and conditions in her collective, claiming that miners who visited the collective were usually shocked and had often commented that the women working there would have an easier, and much better-paid, time working underground. Despite her compelling description of the discomfort of the boiler house, however, the speaker did not elicit any sympathy among the delegation from the *lampovaya*, the outspoken leader of which was scornful. She stood up to oppose a pay rise for the *kotel'naya* and claimed (erroneously) that the conditions there were no worse than those in the *lampovaya*. The incident ended with the issue of rhetorical invitations between the two collectives, each daring the other to come and discover the nature of real work. Thus, while workers, when criticising management, often conjure up the image of 'simple workers' united in their oppression, there is actually very little solidarity between workers outside the confines of their work groups.[18] This partly reflects the strength of

[18] This is confirmed by research at other enterprises. A survey of workers at a Moscow light engineering factory, for example, found that workers were only ready to support members of their own subdivisions in defending their rights (Kabalina *et al.*, 1996: 242)

particularism based on the insular cultures of immediate work collectives, but at the same time it highlights the weakness of workers' identification with each other as workers.[19]

The potential for collective organisation on the basis of the collectivism of the immediate work group is the subject of the next chapter, but before moving on to examine relations at this level it is important to ask why a more general form of workers' solidarity, which is so frequently invoked rhetorically, has not emerged. Why do the *neblatnye* remain as a negatively defined grouping? Why, in concrete moments of conflict, do workers so rarely unite against their managers?

Contested identities

Workers at Taldym mine express their attachment to the collective at every turn, but at the same time the precise focus of their identification is difficult to pin down. At first sight it seems that workers have a strong loyalty to 'their mine' despite all its imperfections. It soon becomes clear, however, that the workforce is deeply divided between the *blatnye* and *neblatnye.* But the *neblatnye* do not stick together when it counts. Meanwhile, although loyalty to the immediate work group is strong, even at this level, as will be seen, collective identification rarely extends into sustained and active mutual defence. In short, it becomes apparent that there is no 'real' site of collectivity, but that the collective identification of workers shifts according to the circumstances. These subjective shifts in perception are possible because workers are relating to a contradictory environment in which they do share some interests in common with mine management, but also feel exploited and cheated by the very same managers. This means that the forms of collectivity that exist, though real enough for the workers at the moment of articulation, are never given any kind of organisational expression. So, as mentioned above, the *neblatnye*, though they definitely feel a community of interests with each other, are really only a notional, negatively defined grouping. Meanwhile, all-mine collectivism takes an alienated form in which all expectations are directed at the figure of the director.

Their shifting sense of collective identification renders workers peculiarly vulnerable to managerial manipulation. When discontent is simmering, managers, whose worst nightmare is independent worker mobilisation, employ the ideal of the united labour collective to full effect. Although in casual conversation identification with the labour collective as a whole does not arouse anything like the passion which workers summon up when they talk about *blat*, 'our mine', does, as argued above, have a special place in their hearts. Managers appeal to this reserve of latent loyalty on a frequent basis. The example of the

[19] What this particular example also reveals is women's lack of solidarity with each other as women.

shift meetings on the eve of the Miners' Day celebrations in 1996 was discussed in the previous chapter. Such meetings also occurred in the era of the previous director: the hostility between the trade union president and the mine director did not prevent them from working together in an effort to contain conflict at the mine. For example, in October 1994 Rosugleprof held a demonstration in Kemerovo in protest at late payment of wages and the government threat to social subsidies channelled through the mines. At a time of rising tension, the trade union at Taldym mine held shift meetings to inform the workers about the protest. Again these meetings, like those in August 1996, were designed to emphasise the community of interest between the mine director and the workers. The trade union president gave a speech in which he explicitly said that his union would be working in co-operation with the mine directors against the government. The mine director, meanwhile, also attended and in his speech gave his blessing to the protest (to which each mine had sent only a handful of representatives in order to avoid any disruption). At the same time he announced a new paternalist initiative: the mine had received some new cars in a barter transaction and it would sell these for cash in order to pay arrears in holiday pay, make a few discretionary payments to those especially in need and buy a new bus. This precisely parallels the strategy of his successor at the August 1996 meetings: the decision to use the cars for the good of the many, rather than distribute them to the few, appeals to exactly the same sentiments as the decision that the office workers would not receive an 'advance' on their wages to spend on Miners' Day.

It is clear that, for the moment at least, such tactics will prove successful at Taldym mine. For all the workers' disdain of the management, the appeal for the labour collective as a whole to struggle against Moscow is very potent and allows the director to structure workers' grievances to suit his own purposes. Meanwhile, the workers, because they have no alternative organisation of their own, lose sight of their own interests. They continue to moan about injustice in the enterprise, but, for the moment, workers' organised action does not go beyond their participation in management-controlled campaigns against the government.[20] Meanwhile, when anger is directed at management it is within a 'one enterprise' framework: workers worry that their present managers are not capable of getting the mine through the crisis. In spite of episodic eruptions when the collective dismisses its leaders with abandon, the leader remains, for good or ill, the embodiment of the collective. It is the chief who is responsible for the good or bad fortune of the collective, and the immediate solution to bad fortune is not to give more reality to the collectivism of the labour collective, but always to elect a new chief.

But why do workers not challenge this vision, even if only on a limited basis

[20] This recalls the events of the 1989 strike. In this case, the workers did begin with their own agenda but the mine directors did their best to retain at least the appearance of a common interest with them, while the regional authorities were able to use the miners' strike as a means of pressing their own demands against Moscow (Clarke *et al.*, 1995: 17–80).

within the mine? While the chapters on trade unions have shown that there are objective reasons for the labour collective to unite against the 'outside world', the interests of workers and managers do, none the less, diverge and workers are well aware of this fact. Why do they not seek to resolve their problems with management through independent organisation?[21] In part, of course, it is because although the Soviet authorities promoted ritual collectivism within enterprises, they suppressed any form of independent collective organisation. But this cannot fully explain why now, in a much freer climate, there is scant sign of burgeoning collective organisation. Fear of the authorities has to some extent been replaced by fear of redundancy, but though this is a hindrance to worker organisation, at Taldym such fear is in no way overwhelming.[22] Workers still have the courage to talk about their individual grievances. Managerial tactics are another factor, but why are they so successful?

The solution to this problem in fact comes back to the system of state paternalism described in the first Chapter. As has been argued above, in the absence of money, relations within the enterprise became highly personalised. Position was all-important, and so were relationships to those in a position of power. The discretionary nature of benefits within the enterprise forced workers to confront those in authority as individuals rather than as members of a collective. So though ordinary workers generally felt themselves to be outside the ranks of the *blatnye*, it was almost impossible for them not to become to some degree ensnared in the endemic special pleading and deal making within the enterprise. As has been shown above, notwithstanding the partial monetisation of the Russian economy, connections still play a crucial role in the everyday life of the enterprise: workers have not been liberated from their dependence on officials by market reform. To challenge the relentless individualisation of relations within the enterprise is very difficult precisely because most workers are

[21] This question will be returned to in more detail in Chapter 8. It is obviously complex because some the important problems confronting workers cannot be resolved within the enterprise, most notably that of wage delays. As one worker put it, 'Neither the brigadier, the shop chief, nor the trade union can do anything about late pay.' Nevertheless, as indicated in the section on *blat*, workers are under the impression that managers and those connected to them are paid before they are. It is therefore legitimate to ask why workers make no collective attempt to deal with this and similar forms of injustice about which they constantly complain.

[22] The 'whip' of the labour market is beginning to exert a greater influence on Russian workers, and fear of dismissal is now sometimes cited by workers as a reason for their lack of protest. (See, for example, Kabalina *et al.*, 1996: 243.) The extent of this should not be exaggerated, however. Unemployment has so far not followed from reform in the automatic fashion that most economic commentators expected, and it does not constitute a sufficient explanation for workers' lack of protest in the present period. Workers at Taldym do claim that managers' habitual response to any complaints is now: 'You don't like it? Then leave.' In fact, however, workers at the mine are very rarely sacked for anything other than drunkenness (and even this has only become firm policy since 1995). Workers may have become more concerned about hanging onto their jobs than they were in the past, but this is not why they fail to organise collectively: they do not even consider collective organisation as a response to their problems. This ingrained individualism is what the following section seeks to explain.

implicated some way or another, and are used to the system.[23] Their instinctive reaction to any problem is to seek an accommodation with the appropriate member of management.[24] Collective action, meanwhile, is an unknown quantity with the potential to destroy predictable, and sometimes profitable, relationships.

This does not only apply to the sort of personal problems taken to the director's surgery discussed in the section on the labour collective. Workers will often make individual approaches to the director on questions of a collective nature. And they often deliberately couch their pleas in personal terms, playing on his pride and appealing to him as a 'human being' to live up to their expectations of a 'good father'.[25] The shop trade union president from the *lampovaya*, for example, was able to trap the director into agreeing to pay for new overalls for her collective through precisely such means, by questioning the truth of his claims that he would love to help if only he had the money to do so. Her account of her encounter also neatly captures the atmosphere of the paternalist enterprise (though there is obviously a gendered aspect to this tale):

> I kept going to the director until he agreed [to give us the overalls]. I was in his office nearly every morning. Finally one day he said 'Do you know what a fuss the chief accountant will make if I sign for you to get this money for new overalls? She told me off only yesterday for signing too much money away.' I said, 'OK, but so that I know that you've really tried, phone her now while I'm here and we'll see what she says.' He said 'Oooh, I like people who deal with me like this – that's the way to get something done.' So he called her. He said, 'You know that I prefer to undress women, but in this case I'm going to do the opposite – I'm going to dress the *lampovaya*.' She [the accountant] said that there was no money at the mine that day, but that I could go and get it on Monday.

[23] Crowley (1997) argues that miners successfully fought against injustice within the enterprise (196). But, as the evidence of the previous sections shows, this is precisely what they have failed to do.

[24] In the past one of the options open to workers was to make an individual complaint to the mine Party committee or the *gorkom*. As has already been discussed in Chapter 2, if the problem was deemed sufficiently serious then the Party would often intervene on the worker's behalf. The evidence from Taldym completely supports Ilyin and Ilyina's (1996) assessment of the role of the Party in the enterprise: 'If at the level of the country as a whole the liquidation of the CPSU was a great leap forward in the democratisation of the country, in enterprises the liquidation of the primary party organisation meant the liquidation of the sole effective form of control of the activity of the administration on the part of the labour collective' (378–9). It should be stressed, however, that the control exercised by 'the labour collective' was not 'collective' in character: generally the Party would be approached by dissatisfied individuals rather than groups and the Party decided whether or not to pursue the complaints 'in the interests of the collective'.

[25] Such appeals are surprisingly potent in the Russian context. It is a matter of pride to be able to provide to petitioners (on a selective basis, of course). The present director, for example, who is currently waging a campaign against drunkenness, was persuaded to give a job as a fitter in a development shop to the alcoholic husband of a woman he had worked with as a young shop chief. The man in question had been sacked from the open cast mine and refused employment everywhere else he had applied. The Taldym director, however, immediately agreed to take him, reportedly with the words, 'A man who can't help a friend is no better than a pig.'

The personal, and especially woman's,[26] touch is frequently resorted to in this way to deal with a whole variety of problems. For example, a similar scene was played out at the end of two separate large-scale mine meetings. After a meeting of brigadiers in October 1994, a brigadier from the technical complex sidled up to the director in the corridor and appealed to him, as a paternalist, to improve the wages of women workers of the technical complex. The director smiled, but promised, and delivered, nothing. Meanwhile, after a meeting of the shareholders' society in June 1995, the brigadier from the *zaryadnoe* approached him, and asked for the 'Kuzbass' bus service, which serves the 'private sector' housing of the settlement, to be reinstated. This woman had a stronger relationship with the director, and her request was cheaper to fulfil. 'Of course, I'll see to it at once' the director replied, smiling, 'I didn't even know it had been stopped.'[27] This culture of personalised pleading obviously strongly militates against the collective organisation of workers.

Despite this, there are instances of spontaneous collective defence within work groups, but these almost always remain on an informal and small-scale footing. At the more general level, workers will sometimes mobilise over a specific issue, usually a management decision which has a negative effect on their pay or working conditions, but such mobilisations are fleeting and do not result in the establishment of lasting organisational structures.[28] Moreover, such activity rarely unites the *neblatnye* as a whole, but is usually confined to one group of workers, in most cases mining brigades. Thus, despite all the subterranean tension which *blat* engenders, within an environment in which access to resources is so crucial the patronage system at the enterprise is neatly self-perpetuating.

Monetisation and informal relations

Is the gradual monetisation of the economy[29] undermining patronage networks and informal relations within the enterprise and ushering in a society

[26] Prostration in front of the paternalist is obviously culturally easier for women than it is for men. Nevertheless, although men do not appeal to the director as a 'big, strong man' in quite the same way as some women workers do, they still take care to cultivate informal relations, relying instead on notions of honour and responsibility to 'the lads'.

[27] In spite of this promise the bus service has now ceased for good. This points up the fact that while such personal appeals may have borne fruit in the past, they are not actually an effective way of dealing with the problems created by transition. Old habits die hard, however, and workers may have to suffer a good deal more disappointment at the hands of imperfect paternalists before they adopt new approaches.

[28] The miners' and female surface-workers' success in their campaigns to change their respective shift systems (the latter of which is discussed in the next chapter) are an example of this.

[29] The monetisation of the Russian economy is an uneven process. The development of market relations means that the rouble does now act as a universal equivalent, but, paradoxically, monetisation has been accompanied by a parallel demonetisation of exchange relations. This reflects the lack of liquidity in the Russian economy resulting from the combination of the 'tight money' poli-

governed by 'legal-rational' criteria? So far there is no evidence of this. Individualised informal relations extended beyond the regulation of access to goods and services, so even were *blat* to lose its significance in this regard, the culture of the enterprise would not change overnight.[30] Moreover, the examples cited above show that while the various waiting lists at the enterprise may be becoming a less salient feature in the reproduction of a culture of individualised relations at the enterprise, for the moment *blat* still plays an important role in the organisation of everyday life.

Informal relations governed all aspects of the life of the Soviet enterprise. The 'untechnological, unduplicable, unreproducible' character of Soviet and post-Soviet production (Alasheev, 1995b: 93) means that line managers are highly reliant on the informal co-operation of workers to keep a rickety system on its feet.[31] Without significant improvements in supply and organisation, it will be difficult to transform this relationship of mutual assistance: for this reason, none of the attempts to transform relations on the shop floor in the Soviet era was successful.

Meanwhile, the culture of punishment for workplace offences was, like the system of distribution, personalised and discretionary. Managers often did not apply immediate sanctions for disciplinary infractions, preferring instead to create a drawn-out dependence in the worker who had been 'let off this once'.[32]

cies of the government and a tax regime which discourages enterprises dealing in cash. Thus, while money can now be freely exchanged for goods, enterprises have increasingly resorted to barter and workers have been forced to live without wages for months at a time. The 1998 collapse of the rouble raises the spectre of non-monetary forms of exchange assuming a new importance.

[30] This conclusion is supported by the sociologists Kabalina *et al.*, who, on the basis of their study of the Moscow light engineering enterprise, argued that 'monetisation penetrates and strengthens the informal relations which . . . maintain production'. In the words of the section chief they quote in their study, 'Labour relations cannot be formalised at all levels, because management simply will not play by the rules, informal relations will be preserved' (Kabalina *et al.*, 1996: 211). Similarly, Sergei Alasheev, in his study of informal relations in the Russian enterprise, concluded, 'If in the past it was important to have acquaintances, connections with Komsomol, Party or trade union bureaucrats (the biggest distributors), these organisations have either vanished or their distributive functions have been curtailed, but connections with the management of the factory, with the shop chief or foreman, with the production structures, remain important' (Alasheev, 1995a: 67).

[31] In the same way, enterprises are very dependent on line managers running around and engaging in all kinds of informal deal to secure the supplies needed by their workers. For a detailed examination of this element of the Soviet production system see Ilyina, 1996a.

[32] For a more detailed discussion of informal relations in Soviet production see Alasheev, 1995a. Specifically on workplace punitive regimes see Alasheev, 1995a: 52–8. The approach of line managers to discipline and punishment is perfectly captured by one of the foremen quoted by Ilyina in her ethnographic investigation of the role of foremen: 'If a fitter has drunk a bit, the most important thing is not to punish him, but to notice the fact and make sure he realises that I have noticed. Who has not had one or two? If he is a good worker, if this is not regular, but a lapse, if he has drunk a bit, then I can cover for him, and he will be grateful to me and will work even better than before, and I will be respected as a tolerant person and as a boss who does not punish people for trifles' (Ilyina, 1996a: 75).

This culture was explicitly mentioned by workers at Taldym as a reason why it was impossible for them to secure improvements in their position. One of the women workers from the *zaryadnoe*, for example, complained bitterly about the high concentration of gas in her depot, and the prevalence of respiratory diseases in the collective. Asked why it was not possible for the workers to press for the installation of a more effective ventilation mechanism, she replied:

> Because managers always have something to hold against you. Either they've seen you reading at work – which is forbidden – or they have seen you going to the canteen. Officially we don't have the right to go to the canteen on our shift system. They know that we go, but then if they need to they can use it against us. Then there's the night shift. When we have done all our work we often have a sleep on the bench here. You never know whether or not you have been seen. This is something else that can be used against you. There is always a reason to punish or even sack you, so it's impossible to open your mouth.

This form of individual dependence will only be strengthened as economic reform progresses, because while in the past workers would not have been sacked for the sorts of offence described above, the experience of enterprises where there have been large-scale redundancies shows that violation of workplace discipline does become more dangerous when decisions need to be taken as to which workers to dismiss.[33] Thus, although fear of the sack is not yet a major determinant of workers' behaviour at Taldym, it may well become more important in the future.

Similarly, transition to a system in which workers receive a straightforward cash wage rather than a series of discretionary benefits in kind alongside their wage will not of itself erode the culture of personalised relations within the enterprise. Payment systems in the transition era will develop on the basis of norms inherited from the Soviet past – and there was always a discretionary element in the payment of wages as well as in payment in kind. Although payment scales were highly formalised and in principle strictly controlled by norms and scales laid down from above, in practice line managers enjoyed a considerable amount of discretion in determining the earnings of individual workers, particularly through the allocation of piece work, the distribution of premiums and bonuses, and the imposition of various fines and penalties.[34] Similarly, at an enterprise level, the wages of particular shops or brigades could be inflated beyond the limits of the official pay system according to the needs of production and of management. This discretionary system is what the trade union president at Taldym tried, and failed, to eradicate. His failure is not at all surprising: discretion is the key to managerial authority within the enterprise.

[33] For ethnographic accounts of the way in which staff reductions have been carried out in other industries see Metalina, 1996, and Kozina, 1996.

[34] For examples of this see Vedeneeva, 1995: 224–39, and Donova, 1996: 41–62. Ilyina also notes that foremen are able to determine the coefficient of labour participation (KTU) as a way as rewarding or punishing workers, and can encourage workers through use of the so-called 'foreman's fund' as well (Ilyina, 1996a: 76–7).

This chapter has provided part of the explanation for the lack of independent workers' organisation, concentrating on the shifting nature of workers' collective identification and the culture of informal relations within the enterprise, which, it has been argued, promotes individual dependence and discourages collective organisation. This still leaves one question unanswered: are there any mechanisms through which conflicts of a collective nature can be resolved within the enterprise? As has been seen above, workers are able to resolve some of their problems through individual approaches to members of the administration. But questions of a collective nature generally cannot be resolved in this way, as the failure of the individual pleas for a pay rise for the technical complex and the restoration of the Kuzbass bus service reveal. So how do immediate work collectives deal with their problems? Are there any channels through which they are able to express their grievances? And if workers do collectively defend their interests within work groups, could this form the basis for wider workers' organisation? The next chapter will consider these questions, focusing on the nature of collectivism and collective action at the level of the immediate work groups.

7

The lampovaya*: the stunted collectivism of the immediate work group*

> The *lampovaya* is the heart of the mine. (Quotation from the former forewoman, Anna Petrovna's, retirement album)

To assess the ability of the workers to give a more substantial reality to collectivism within the enterprise, we have to turn from consideration of the collective as a whole to analyse the immediate work collective. The collectivism of the immediate work groups, like that of the labour collective, has a symbolic dimension, but it has a more tangible reality in the particular social relations formed at the level of the work group. If a new form of collectivism is to emerge it is most likely to be based in the first instance on the collectivity of work groups. The next section of this chapter therefore focuses on the nature of workers' relations to their immediate work collectives. Then, on the basis of a case study of one such collective, the *lampovaya*, the chapter goes on to chart the development of social relations in these collectives from the late *perestroika* period onwards. It argues that while the collapse of the Party and the erosion of discipline at the mine allowed the work collectives more autonomy, workers have proved unable to develop the collectivist potential immanent within immediate work collectives. This is related to the alienated form of collectivism fostered within the enterprise, which means that, even at the level of the work group, workers' aspirations are focused on those they perceive as their patrons.

'My collective – my second family'

The immediate work collective was the point at which the individual's integration into Soviet society was monitored and regulated, but it was also a focus of sociability where workers spent half their lives together. The effectiveness of the work collective as a locus of social integration and control depended on the fact that individuals were attached to their work collectives. This section describes the relations of workers to their collectives and highlights the centrality of the collective in their lives.

The precise significance accorded to the collective by workers differed from that prescribed in official discourse, but the idea that the work collective was a crucial reference point was common to both. This is true for both male and female workers. Men and women do not relate to their collectives – which in the mine are almost exclusively either all-male or all-female – in precisely the same way, but both attach similar levels of importance to their lives within the collective.[1] The same point can be made about different categories of workers, such as male surface-workers and miners.

The following discussion of relations to the collective draws on interviews with male and female workers at Taldym conducted in the years 1994–6. The quotations used in this section illustrate a particular kind of attachment to the collective which still exists, but it is important to point out that there is an increasing tendency for workers to note that their collectives are becoming less close, that life at the mine is becoming 'less interesting', and that work is becoming less important to them because 'they don't pay us'. This was especially marked in August 1996, a time at which the mineworkers had not been paid for three months. The following section thus isolates a set of attitudes which appear to be being transformed in the present stage of transition, though they none the less still inform workers' normative expectations: work and the collective may be becoming less important, but this is far from a welcome development. Indeed, the pervasive sense of loss induced by the waning of this form of communal sociability fuels the longing for security and protection which characterises contemporary Russian political life. This development is one of the themes of the next chapter.

Both male and female workers describe the collective as a welcome retreat from the pressures of home, though since men and women face different domestic pressures, they tend to look for a different kind of relief at work. For women, there are two main themes. First, women workers often claim that their work within the collective is less arduous than their work at home. At home they feel isolated and under constant pressure, while at work their load is lightened by companionship. Since work at home not only involves traditional 'women's work' such as washing and cooking, but also includes fruit and vegetable production and often animal husbandry as well, it is often at least as physically exacting as work at the mine, if not more so. This was the view of a brigadier from the *zaryadnoe*, who keeps cows, pigs and chickens:

> There are breaks at work when you can rest. You can't rest at home. There is always something to be done. I have the cows to look after. My mother helps me a lot but there is still a great deal to be done. The home has to be in order. . . . Work is hard, but not always. Perhaps it's because you are in a collective that the time goes quicker. It seems that soon after you arrive it's time to leave. You can chat and you don't notice the time going by. At home you work on your own.

[1] For more details on women's attitudes to work see Ashwin and Bowers, 1997.

Thus, rather than home being viewed as a refuge from work, for women the collective in some respects acts as a haven from the pressures of running a household.[2]

Second, women describe the collective as a social focus and vital source of emotional support. This can be seen, for example, in women's reappropriation of the Soviet era expression 'Off to work, like to a holiday' (*na rabotu kak na prazdnik*), which, along with expressions such as 'We pretend to work and they pretend to pay us', made joking reference to the supposedly endemic slacking among Soviet workers. This expression cropped up several times in interviews with women workers from a variety of collectives, but it was not employed ironically. Instead it was both posited as an ideal and used descriptively. One worker from the *lampovaya*, for example, talking about what she felt was the wonderful atmosphere that had prevailed in her collective before the forewoman had decided to break up established shift teams, claimed:

> I used to look forward to going to work – I went to work *kak na prazdnik*. If I had any problems I used to get to work and tell the girls about it and they'd say, 'Don't worry Lyuda it'll all work out' and I already used to feel better. We all used to help each other. For example, we used to fight over the floor cloth: I'd say, 'Look you have a rest, I'll do the floor today' and they'd say, 'No Lyuda, you're always doing it, let us do it.'

In their emphasis on the value of their collectives, women have adapted communist ideas of the past in the light of their own experience: they treat the idea of the collective with a reverence that the communist authorities would have found very pleasing, but in place of the rather sterile idea of the collective as a cell working towards the construction of the communist future, for women the collective is a vital support network. The seriousness with which women approach the role of the collective can clearly be seen in the following comments made by the brigadier of the *zaryadnoe*. She explained how a worker had recently joined the collective, and, as was the brigadier's custom, while explaining the job to the new worker she had taken the opportunity to ask her about her family background and so on. The young worker resisted answering her questions and finally said, 'Don't pry into my soul.' At this point, the brigadier felt compelled to explain the meaning of the collective to the new worker: 'I'm sorry, but we all know each others' problems here. The collective is your second

[2] A number of women argued that work at the mine was actually a defence against housework: because they worked at the mine (usually on twelve-hour shifts) and were not always available, their partners were forced to take more domestic responsibility. Most of the husbands of mineworkers with children had, for example, been forced to develop their culinary skills at least to the level where they could feed the children when their wives were on shift. One worker from the *lampovaya* made exactly this point. Her husband is a face-worker in one of the most successful shops at the mine and has enough money to support his family. He had been trying to persuade his wife to give up work, but she had no intention of doing so. She liked the fact that her husband gave her some help with her housework, and claimed that this partnership would be instantly dissolved if she agreed to give up work: 'You're at home all day, do it yourself, that's what our men say.'

family. You come to work and you can express [*vyskazat'*] your feelings, talk about your problems and then you'll feel better. That is how it should be.' The brigadier was genuinely perturbed by the attitude of the new, 'closed' member of the collective and she was determined that she should learn the proper place of work in life. As this example shows, feelings about the importance of the collective do not just reflect daily experience, they also contain a normative component: this is what work should be about.

The idea of the collective as a 'second family' constantly recurred in interviews with female workers.[3] Indeed, it functions as a buttress to the actual family, which is usually portrayed as the cause of distress. The following formulation of a worker from the *lampovaya* is typical: 'We are like a family. We can discuss personal problems together, for example, if someone has a problem with her husband. We give advice. Everyone has someone in the collective whom they trust.' Meanwhile, the collective also provides stimulation which is often lacking in the home. The latter point is well captured in the explanation of this retired worker from the *lampovaya* as to why she wanted to return to work:

> The collective is your second family. I, for example, do not need to work for the money – my husband and son are miners. But I will look for work in the winter because I need to be in a collective. I need to be with people. Housework gets very boring. And your family, well you can get bored of them too if you don't see anyone else. Human beings need to socialise.

Male workers do not express their attachment to the collective in these terms and would only very occasionally refer to the collective as a family. This is partly because the prevailing norms of masculine behaviour mean that it is less appropriate for men to describe their relations using 'feminine' family imagery, but it also reflects the fact that the nature of the respite provided by male collectives is different: while female collectives tend to have a confessional, confiding culture, for men the collective offers light relief. The difference is well captured by the comments of a group of fitters from one of the in-mine transport shops:

> FITTER 1: How could we survive now without the collective?
>
> FITTER 2: You come to work and the collective raises your spirits. . . . we're a cheerful collective.
>
> FITTER 3: Yes, we've got our own 'circus' here, we don't need to go and pay for it. He's on leave at the moment, although our clown's still here.

As mentioned in Chapter 2, it is common for such collectives to go drinking together after work a few times a month,[4] and many workers also go fishing

[3] Such ideas are by no means exclusive to women working in mining. A recent study of a Samara chocolate factory with a predominantly female labour force revealed exactly the same attitudes: the conference paper in which they were analysed was entitled 'My factory – my home' (Romanov, 1996).

[4] Many male workers noted, however, that such impromptu gatherings were becoming less common. Explanations for this included the lack of money for vodka caused by wage delays and the fact that work on allotments was taking up so much of their time.

with work-mates in summer. A significant part of the pleasure of these encounters derives from the all-male company, which again provides a haven from what male workers tend to portray as the female-dominated domestic world. The value placed on the all-male company of the work collective is well captured in the bemused response of the fitters quoted above to a question about why they didn't drink with women:

> FITTER 1: How could we? It's not possible. It's after work and we're together and she's got things to do at home.
>
> FITTER 2: We're an exclusively male collective. The only time we drink with women is on holidays and birthdays. They drive us out [of the home].

Their collectives offer both male and female workers companionship and an important forum of self-expression, although, as indicated above, the nature of the interaction within the collectives differs. Partly as a consequence of the different functions of the collective in the lives of men and women, a notable distinction between the two groups is that male collectives tend to socialise together outside work time – clowning is even more enjoyable with the addition of vodka – while women workers tend to have separate networks of friends outside work.

Another distinction in the attitudes of the two groups is that men, unlike women, do not see work at the mine as less onerous than the work they perform at home, which includes tasks such as tending the family plots, chopping wood, and, in the case of those with livestock, hay-making in summer. In the case of underground workers, the reasons for this are self-explanatory. More generally this difference can be explained by the nature of the sexual division of labour in the home. Women not only have more work, they also feel that work at the mine is more meaningful and socially valuable than cooking and housework, the prevailing attitude to which is well captured in the comment of one woman worker, 'You cook a meal, it gets eaten and then what have you got to show for it?' Meanwhile, the productive and creative nature of work on family plots, which constitutes the bulk of the male contribution to the household,[5] means that both men and women tend to view it not only as a necessity, but also as a hobby.

Both men and women workers have a strong sense of identification with the collective which gives them both support and a sense of meaning in life. The importance attached to the co-operative culture at work, and the value of the relations built up over time, are well captured by the warmth and feeling with which the miner quoted below spoke of the (model and highly productive) brigade to which he had belonged for nearly his whole working life. The brigade had been transferred *en masse* from a nearby mine to Taldym, where it worked for several years before being disbanded. Recently, however, its members had been reunited when the (locally famous) brigadier agreed to return to work

[5] Although men put in no more time working on the family plots than do women (Varshavskaya, 1998).

after a long illness – a history which in itself says a good deal about the strength of relations within the collective. The miner portrayed a solidaristic culture of mutual support and understanding:

> Once they divided us up. I don't know why. It was simply stupid. . . . It seems to me that when a collective's friendly, as they say tight-knit [*skolochennyi*], where everyone knows each other, you work better because you understand each other in half a word, half a look, even. . . . We're very close [*splochennyi*], not quite as much as before . . . but before we were very friendly. The collective alone decided everything. For example, if someone worked with us in a low skill category, and the collective decided it wanted to we'd go to the brigadier and say, 'This person works with us and like us, and he should be paid the same as us, regardless of his category.'

The attachment to both the social reality and the idea of the collective which comes across in this quotation and those of the women workers cited above is a distinctive feature of Russian working life. But does the collectivity of the work group provide a basis for independent workers' organisation? The rest of this chapter attempts to answer this question through an analysis of the *lampovaya.*

The work of the lampovaya

> Although there are individual exceptions, the same situation exists at *every* enterprise. It [corruption] starts in the Kremlin and goes down to the mine level. You find it at the smallest kindergarten, at every enterprise – everywhere in Russia it's the same. (Anna Petrovna, former forewoman of the *lampovaya*)

The *lampovaya* is a small collective of only thirty-two workers. This chapter argues, however, that events in the *lampovaya* exemplify in microcosm processes occurring both within the mine and more widely in other Russian enterprises. For, while workers feel a strong identification with their immediate work collectives, they reproduce within them the same hierarchical and paternalistic form of relations which exist within the mine (and indeed the coal industry) as a whole.[6]

The work of the *lampovaya* consists of giving out and collecting the miners' lamps;[7] checking the miners' attendance and ensuring they return

[6] The following analysis and the narrative on which it is based have been chosen because they starkly reveal a dynamic which could be observed in collectives throughout the mine. This claim is substantiated at the end of the chapter in a section which discusses the form the same process takes in other collectives.

[7] Until the spring of 1996 the miners collected their lamps themselves, dropping a token with their number on it into the boxes of the different shops lined up along the 'front desk' of the *lampovaya* as they left. Then the *lampovaya* was 'closed' and the women were required to hand the miners their lamps and return them to the shelves themselves. Management introduced this change –

from the pit; overseeing the recharging of the lamp batteries; and testing the lamps on a daily basis to ensure their methane-detectors are working.[8] The workers are also expected to clean the *lampovaya* once a shift.[9] It is the responsibility of the workers of the *lampovaya* to notify the mine duty officer if any miners have failed to return from the pit. They are also supposed to mark down in the *grafik* books any miners who return before the end of the shift, so that they can be punished, but in practice the women often exhibit discretionary blindness when it comes to spotting marginally early returnees.[10]

The *lampovaya* is an autonomous unit within the mine, managed by a forewoman (*master*), although the collective formally comes under the jurisdiction of the chief of the ventilation and technical safety shop. The collective is divided into four brigades of seven workers. The brigades are in effect shift teams,

which is unpopular among the women because it means more work and among miners because it means they often end up waiting in a queue – because they deemed that too many lamps were going missing. The women discussed requesting a time-and-motion study to determine whether their wages should be raised as a result of the change, but they were unable to reach agreement on the issue. Eventually, one of the most influential members of the collective, Lena the brigadier, put an end to the discussion. She argued that if they were not united on the question there was no point in requesting the study. During such a study it was important for everyone to appear rushed off their feet and the dissenters could not be relied upon to put on the required show. In this situation the study could even result in their wages being reduced.

[8] From the summer of 1995, all the miners have been provided with SMS lamps, which signal the presence of methane by flashing at a speed which indicates the concentration of the gas. These lamps have been introduced over a number of years, and now that all miners at Taldym are equipped with them it is far harder for them to work when there is a danger of explosions: the flashing makes it impossible for them to see properly. The mechanism of the lamps relies on chemical change in one of the components in the electrical circuit of the lamp, which causes the light to flash. After continuous low exposure to methane, however, chemical change means that the component ceases to be effective. Since there can be no precise 'expiry date' in such a process the lamps are checked by the women before each shift. (In UK mines miners themselves check their lamps before they go on shift.)

[9] The somewhat obsessive cleaning of the *lampovaya* appears to be an example of the work-creation ethic of Soviet enterprises mentioned in earlier chapters. It is an official part of the collective's duties, but this thankless work is undone in an instant by the comings and goings of dusty miners.

[10] The present forewoman, in contrast to her predecessor, never lets miners off for finishing work early, and her workers are unable to do so if she or any other members of management are around. Many workers will, however, let miners off if they get the chance: a high proportion of them are married to miners. The vigilant comrades of the collective are disliked both by their colleagues and by the miners, as the following quotation illustrates: 'If a miner comes up from the pit early – let's say half an hour or ten minutes early – I am supposed to note it down, and then they reduce his monthly premium a bit. Well, I can't do it. I always think of his family – they need that money. I feel sorry for them. And the miner – perhaps he's got to get off early because of his family. We always want them to come home early. . . . There are others, though, who always report the miners. Nadia's that sort of woman. They are all afraid of her. With me, they know I'm a softy. They come up and say, "Look, I've got to get off a bit early today". They know I'll never report them.'

working according to a twelve-hour shift system. In addition to the brigade members, the collective comprises one *tabel'shchitsa* (time-keeper), responsible for keeping records of miners' annual leave, sick leave and attendance and undertaking the associated paperwork, and two fitters responsible for repairing defective lamps. These members of the collective work a five-day week, eight until four, as does the forewoman.

Each brigade is led by a brigadier, responsible for ensuring that the shift fulfils the required amount of work. Unlike in some other collectives the brigadiers are not elected, but are appointed by the forewoman. There is actually no competition for the place of brigadier, because it is she who is responsible for any mistakes made by her workers. For this reason, the forewoman tends to rotate the position among the most senior members of the collective. As one brigadier commented:

> Even if everyone who should be on shift is at work, I still have to answer for any mistakes that are made. If I don't notice and correct a mistake in time, it is counted as my mistake. . . . No one wants to be brigadier, so we only do the job for a year and then change over. No one wants the responsibility.

Meanwhile, within each brigade there is a *tabel'shchitsa* who is responsible for filling in the attendance of the miners into the *grafik* record books of each shop on a daily basis. This task is also in principle rotated within the brigades, but the forewoman decides who should do it, and in practice some workers are chosen to do this job far more often than others.

The forewoman is responsible for ensuring that the work of the collective is properly implemented – that the lamps are recharged and function for the correct amount of time; that the shop chiefs receive all the information they require; that any miners who have failed to turn up for work are contacted and so on. Miners also have to have their requests for annual leave signed by the forewoman of the *lampovaya*: the signature can only be obtained if they do not have an unexplained absence (*progul*) recorded against them.[11] The forewoman assigns duties (the *naryad*) to the brigades and makes sure that they fulfil them; she determines the work roster of the collective. Essentially, she is the immediate line manager of the workers of the *lampovaya*, but, because of the autonomous status of the collective, she has slightly more managerial responsibility than foremen and women in other collectives.

[11] The forewoman exercises this power in the imperious manner characteristic of female administrative staff at the mine. On one occasion she discovered a *progul* in the records of a miner who had come to her for a signature, and her tone immediately changed from grudging politeness to open rudeness. She shifted directly from the polite 'you' form (*vy*) to the familiar *ty* form, and refused to make any enquiries as to whether the records were correct or not. This form of humiliation is a highly effective way of disempowering workers. It is, however, also a major cause of resentment.

'The collective can do nothing': the defeat of the superwomen of the *lampovaya*[12]

The *lampovshchitsy* are committed to their collective, but do the social relations of the work group provide a basis for self-organisation? And can the collective defend itself? The following section explores these questions by examining the development of relations in the *lampovaya* from the late Gorbachev era.

The immediate work collective has always been an important arena for self-expression, though in the communist period this was balanced by its function as a locus of control. But the change in the political landscape in the late *perestroika* era created a space in which the emancipatory aspects of the communal life of the collective could potentially develop. After 1989 much of the disciplinary apparatus was dismantled: the draconian disciplinary code which applied to the mines was suspended in 1989, the Party and the state security bodies have since been removed from enterprises, managerial authority has been substantially undermined, and workers have acquired the ability to remove their managers. The combined effect of these developments was to shift the balance of power in favour of the workers, allowing them to increase their collective control over the production process and to order the lives of their collectives in a way which suited them. But this flowering of collectivism also had its limitations, the nature of which became more apparent as the optimism generated by Gorbachevian reform dissipated in the cold light of the Yeltsinite new dawn. This stunted growth of collectivism can be clearly illustrated by an account of the development of work relations in the *lampovaya* during this period.

The years 1988–92 represented the high point of collectivist development within the *lampovaya*. In 1988, the *lampovaya* became the first collective in the mine to remove its line manager by democratic vote: as the eventual successor to the deposed forewoman proudly put it, 'Democracy came first to the *lampovaya*.'[13] The forewoman of the collective, Daria Nikolaevna, was voted out of office after an attempt to disrupt the stability of work relations by introducing a regime of three-monthly changes in shift teams. As one of the brigadiers from the collective reported: 'She first of all broke up all the shift teams, then terrorised everyone; . . . she did everything awful she could to the collective, and annoyed everyone so much that we simply, quite simply, got rid of her.'

After a brief spell with another forewoman who was deemed unsatisfactory, the collective elected a former teacher, Anna Petrovna, whose four years

[12] One of the many verses the collective composed for Anna Petrovna's retirement album read, '*Lampovaya prosto klass, superzhenshchiny u nas*': 'In the lamp room we're dead cool, superwomen one and all.'

[13] After the 1989 strike, voting unpopular managers out of office soon became common practice throughout the mine. As one shop chief lamented in an interview, 'After the strikes it even got to the point where it was obligatory to change managers – business-like or not – hands up, let's have a different one.'

in office are remembered as a golden age within the collective. In 1992 Anna Petrovna resigned as forewoman on grounds of ill-health.[14] The mine director reappointed Daria Nikolaevna to the post and the collective did not resist this, on the grounds that she had 'learned her lesson' and was no longer a threat. This proved at first to be the case and the working practices developed in the Anna Petrovna era persisted for some time.

After 1988, the workers of the *lampovaya* were able to create a highly conducive working environment for themselves. One of the most important aspects of this improvement was a development not restricted to their collective. The shift in the balance of power in favour of the workers after 1989 meant that the 'women's collectives' at the mine were able to secure a change in their *grafik* from an eight-hour to a twelve-hour regime. Thus, instead of a gruelling regime of three days on a first shift, followed by three days on the second shift and then three on the night shift, each three days only separated by one day off, the women now work one twelve-hour day followed by a twelve-hour night shift with two days off in between each stint on duty. The campaign for the change in *grafik* represents the most significant achievement of collective action among women workers at the mine.[15] It is cited by all the women working according to this *grafik* as a positive aspect of their work – not only does it allow women conveniently to combine their home and work lives, it also weakens managerial control because line managers often work according to different shift systems.

The change in forewoman and *grafik* allowed the *lampovshchitsy* a great deal more autonomy over the organisation of their work. Since the forewoman only works a five-day week and the brigades work on the continuous twelve-hour shift system, for the majority of the week the brigadiers are in charge of the work of their shift teams. And since a brigadier, as shown above, is just 'one of the girls', this means that for the majority of the time the collective is effectively self-managed. Workers were thus able to establish their own informal norms within their brigades and arranged their work in a way that suited them. As one of them reported of her former shift team:

> Lena was the senior *lampovshchitsa* on our shift and she was understanding. For example, if I had to get home early for some reason, I could ask her and

[14] Anna Petrovna claimed that in 1992 she had been to see a witch who had told her that someone was after her place and had cursed her with ill health: a shiver went down her spine and she said she knew instantly that Daria Nikolaevna was the author of the curse. She decided to resign, although Daria Nikolaevna tried to talk her out of it, saying that the collective wanted her to remain as forewoman. Anna Petrovna was adamant, however, and said 'Look Dasha, the craze for democracy is over now; the director will appoint you as forewoman and that will be that.' After Daria Nikolaevna's appointment, Anna claimed that the problems with her health ended.

[15] The twelve-hour shift system was introduced on the initiative of the women workers of the *kotel'naya* who, in a period between shop chiefs, simply took a unilateral decision to reorganise their working hours. Their example was taken up by the other women's collectives, who persuaded the trade union to pursue their case. As mentioned earlier, the eight-hour shift for underground miners at Taldym was a similar post-1989 victory for the miners.

> she'd let me go – without, of course, Daria Nikolaevna knowing anything about it. And on the night shift we used to take it in turns to sleep. Of course, we'd have been punished if anyone had seen us – but we used to get all the work done.

Thus, not only was the *lampovaya* a close collective, it also seemed that in the period after 1988 the collective had developed a capacity for self-organisation which might provide a basis for furthering the workers' own interests. Later developments, however, starkly revealed the limits of the form of collectivism developed within the *lampovaya* in the Anna Petrovna era.

In January 1995, Daria Nikolaevna suddenly announced that she was breaking up the established shift teams, the offence for which she had been removed in 1988. Women who had been working together in brigades for years were suddenly divided up and arranged into new teams, without even being consulted as to whom they wanted to work with. One worker reported:

> When Daria Nikolaevna said she was going to break up the brigades everyone pleaded with her not to do it. She just said 'we'll see' and then went ahead and did it anyway. I think she didn't like the fact that we all got on so well and enjoyed work – I think she thought that it meant we weren't working properly. I don't agree with her. I think if you're happy you work better.

This action had a dramatic impact on the collective: the atmosphere of gloom was palpable on my return to the mine in June 1995. 'We've fallen out with each other' (*my razdruzhilis'*), one of them explained. The workers reported that everything had changed; nothing, it seemed, was the same, even down to the buns of the *stolovaya*, which were formerly a staple snack in the *lampovaya* before some mysterious but spectacular de-skilling afflicted the mine's kitchen staff. The cosy companionship of the old brigades had been destroyed, and along with it the autonomy which the shift teams had previously enjoyed. For example, previously the women had cooked and eaten lunch and supper together in the cubby hole behind the lamp racks or in the 'laboratory' where the functioning of lamps was checked, away from the eyes and ears of the forewoman. Under the new regime, Daria Nikolaevna led a melancholy luncheon party to the *stolovaya* every day. Moreover, because the brigades were not composed of workers who trusted each other, they were unable to organise their work informally as they had done in the past. One brigade in particular was deeply unhappy: they had ended up with the 'good communist'[16] of the collective, who could be relied upon to report any irregularities to the forewoman. As one of the brigadiers remarked of this worker and another close family friend of the forewoman: 'There are tell-tales in the collective who tell Dasha [Daria Nikolaevna] everything. I suppose every collective has such people. No one wants them on their shift – because a shift will always have something it wants to hide from the forewoman.' Whereas these workers had in the past been kept under control by strong brigadiers and workers who knew them inside out, in

[16] She had in the past been an active member of the Communist Party.

their new teams they created enormous tension. One worker even claimed she had taken time off sick to avoid the 'good communist'.

Thus the forewoman had dramatically increased her power over the collective. As one of the brigadiers memorably declared, 'I think Dasha's a vampire. She draws energy from us when we're all quarrelling. It suits her far better when we're not getting on with each other.' The forewoman again reorganised the shift teams at the end of June 1995 without consultation, and announced that they would be reorganised every three months thereafter. Her plan ensured that most members of the collective would eventually have to serve alongside the 'good communist', and that it would be extremely difficult for the workers to re-establish informal working arrangements: the collective had been broken.

There was no collective response to the forewoman's action. A few brave individuals remonstrated with her, but this was as far as the protest went. The collective was awash with talk about replacing the forewoman, but no one was prepared to put herself forward as an alternative candidate. Even when insult was added to injury with the announcement that the shift teams would be reorganised regularly, the workers did not foresee themselves resisting. Asked what would happen if the forewoman went ahead with her plans, one of the older and more outspoken members of the collective replied, 'Nothing will happen. Everyone moans between themselves, but there are very few people who will say it to her face.'

Why had the seemingly well-organised women of the *lampovaya* not been able to resist this assault on their collective? The answer to this question reveals the limitations of the development of collectivism after 1988, as well as the vulnerability of work collectives in the face of managerial encroachments connected with restructuring.[17] First, although in the heady days of the late Gorbachev era workers were able to achieve many improvements in their position at work, this was strongly conditioned by the vulnerability of managers in this period as they accustomed themselves to life without the Party. The *grafik* campaign is a good example of this. It *was* a gain for the women workers at the mine, but it was one which was achieved without a high degree of managerial resistance. Moreover, it did not lead to any lasting organisation among women at the mine. Similarly, once the *lampovshchitsy* had secured what they saw as a desirable amount of autonomy, they were not concerned to build on the incipient organisation they had developed when they first took the decision to remove Daria Nikolaevna. This meant that as management has begun to reassert its control over the enterprise (a process which is continuing but not yet complete), workers have become increasingly vulnerable. The changing

[17] Even though the actions of the forewoman of the *lampovaya* were not specifically connected to restructuring, her action did have the effect of 'softening up' the collective. Any future changes desired by mine management will now be far easier to push through: the lack of cohesion in the collective is well illustrated by the failure of its members to decide on a response to the 'closing' of the *lampovaya*.

possibilities of self-determination within collectives in the 1980s and 1990s are well captured by the brigadier Lena's account of the period:

> When *perestroika* and *glasnost'* began it was easier to change [the forewoman]. Now there isn't any *glasnost'*, there's nothing. Now the managers decide everything again. But before *perestroika* it was absolutely impossible. Gorbachev gave us some rights. . . . He allowed the collective to decide everything. Now there's very little that's decided by the collective.[18]

What is notable about this account is the passive role it (accurately) accords to workers in the process: rights are 'given' and taken away, rather than fought for or defended.

A major reason for the failure to develop the organisational potential immanent within work collectives lies in the structure of the traditional Soviet enterprise, in which the line manager, as well as acting as a representative of management, also represents the interests of her collective to the enterprise administration. The role of the lowest level of mine management is to safeguard the rights of immediate work collectives; to get the best deal for 'their' workers, in the same way as the role of the mine director is to get the best deal for 'his' mine. Rather than defending themselves against the lowest level of mine management, therefore, workers would normally expect to be defended by this 'representative' of the collective. In the Anna Petrovna era, this is exactly what had occurred. Anna Petrovna saw it as her role to stand up for the *lampovaya*:

> When I was forewoman – I am the kind of person who can't bear injustice – and I always fought for justice, always spoke the truth. I got on well with the shop chiefs. But [not] with the *kontory* [office workers] and the director. I complained if we didn't have normal conditions to work in – so that we had some ventilation and so on. Dasha never does that – she won't even ask for a lick of paint to make the place look a bit better. I was always going to the director, to the chief engineer; I was always speaking up for the workers, because they worked well, but they would swear at them for any tiny mistake. I don't know why but none of the directors before . . . have been fond of the *lampovaya*.

Workers confirmed that, in contrast to Daria Nikolaevna, Anna Petrovna had stood up for their rights:

> When Anna Petrovna was forewoman she defended our interests. If we needed something we could simply gather and talk with her either in the shop trade union committee, or just between shifts. She would go to the trade union, the economists, the director, with questions, with demands, saying that they needed to do something for us, that the collective was demanding or asking for something that needed to be done.

[18] This account was largely confirmed by interviews with workers from other collectives, although elite workers such as miners in the famous brigade mentioned above felt that even in the communist era they had the possibility of influencing the director or mine Party secretary. They would, however, only have been able to resolve a dispute with their line manager through recourse to a higher authority – an open democratic rejection of a shop chief was not possible.

> Daria Nikolaevna does not go – she doesn't want to draw attention to herself in the eyes of the president [the mine director], or the eyes of anyone else. She doesn't want anyone to notice and say that she's going there and asking for something or other. She doesn't want it. She wants everything to be quiet and peaceful so that no one notices her and she can get on with her work.

This reveals why, having elected Anna Petrovna, the *lampovshchitsy* considered that there was no cause for further organisation: in place of a trade union representative, she would look after their interests. The problems with this perception were only revealed when Daria Nikolaevna redefined her role within the collective from representative to manager.

Nevertheless, the instinctive reaction of workers to the unpleasant turn in the fortune of their collective under Daria Nikolaevna was still to consider a change in forewoman. The other alternative was to appeal to the mine director for protection, but at this stage he was not prepared to give it.[19] The women did not feel able to resist the change while Daria Nikolaevna remained as forewoman: an alternative figurehead was felt to be a prerequisite for any kind of reaction. Just as the solution to the mine's problems is seen as being the election of a benevolent paternalist, so the solution to the *lampovaya's* problems was seen as being the election of a better 'mother' to the collective. And since the self-organisation of workers does not extend to collective defence this estimation is, in the immediate term, perfectly accurate.

Contested collectivism

Although the *lampovshchitsy* may not be capable of defending themselves against their forewoman they had, as described above, carved out a comfortable communal life for themselves after 1988. Collectivism within the *lampovaya* is limited, but it is not non-existent. The workers do feel attached to the co-workers with whom they have built up co-operative working relations, with whom they have habitually shared meal times, and with whom they have concealed minor infractions of work discipline from the forewoman. It was precisely this form of collectivity which the forewoman broke up when, as the disgruntled *lampovshchitsy* put it, she 'scattered' (*raskidyvala*) the established shift teams.

Daria Nikolaevna herself claimed that she had broken up the shift teams in order to increase the cohesiveness of the collective, saying:

> I did it so that the shifts would all be friendly. So that the girls would all know each other. I did it for the sake of unity [*splochennost'*]. If you work on one shift

[19] Just before he was voted out of office at the end of 1995 the director did intervene on behalf of the *lampovshchitsy:* at the weekly meeting of shop chiefs he publicly criticised the forewoman for upsetting her workers. After this an uneasy truce emerged in the *lampovaya.* Again, however, the workers had not been able to protect themselves – they had to rely on a managerial benefactor.

> with the same people forever, you get sick of it. You could work for years on the same shift and not get to know the other members of the collective. You could pass them on the street and not even say hello. To work as a collective, everyone has to know each other. Anyway, this is my personal opinion.

The forewoman here asserted an abstract view of the collective, and a formal view of unity: not a cohesiveness born out of informal, co-operative relations, but a putative unity forced on the collective from above. Her attack on the collectivity of the brigades is reminiscent of the Soviet antipathy to what in Soviet-speak were referred to as 'false collectives', that is, from the official communist point of view, 'sub-groups that might potentially challenge the supremacy of the *kollektiv* from within, or the devolution of the *kollektiv* into an in-group that does not aspire to support broader social values and pursues only narrow group interests' (Kharkhordin, 1996: 30). The trouble with 'false collectives' from the point of view of the authorities is that they command more loyalty than the so-called 'true collectives' which are subordinate to official control. The forewoman thus 'scattered' the 'false' collectivity of the shift teams, turning the collective into a mere aggregate of unhappy workers, who were only 'unified' in their common subjection to her will. This was quite clear to the workers, who had accurately predicted what the forewoman would say if asked why she had broken up the old shift teams. Their view of her concept of *splochennost'* was '*eto konechno chepukha*': 'it's rubbish of course'.

The ease with which she accomplished her aims, however, reveals the vulnerability of the limited form of collectivism which existed within the brigades of the *lampovaya*. Within the old shift teams, workers had built up relations of trust with each other over a long period and had been capable of ensuring that work was organised in a way that suited them. But the relations which developed were based on co-operation over work: the brigades were work groups, not defensive alliances. This is highlighted by the fact that although the brigadiers could use their discretion in ways which benefited their brigades, when the forewoman decided to break up the shift teams, the teams themselves were not able to put up any resistance. Covering up for each other is a very different thing from acting as a group to defend a collective interest. So, while collective co-operation does exist at the level of the brigade, it barely extends beyond the day-to-day detail of organising work. Workers express a loyalty to 'the collective' – the *lampovaya* – but this form of collectivity is only realised through the medium of the line manager. The collective is not able to represent itself: it can neither obtain resources from the mine administration without the good offices of the line manager, nor defend itself against turn-coat 'representatives'. In the face of a *volte-face* such as that committed by Daria Nikolaevna workers simply look for an alternative figurehead to defend them. In this sense, 'the collective' amounts to little more than an aggregate of individual supplicants. Thus, although the action of Daria Nikolaevna can at one level be seen as merely her whim, it does strikingly reveal the vulnerability of workers to authoritarian assaults from the very matriarchs and patriarchs to whom they look for their salvation: they do not have the organisational resources required for systematic resistance.

Informal representation by line managers

How typical is the *lampovaya*? Are other collectives, in particular mining brigades, more robust and able to defend themselves? The argument put forward here is that while mining brigades may be in a stronger position when it comes to obtaining resources within the mine, they are equally dependent on their line managers to press their case. The director is expected to defend the interests of the mine, shop chiefs try to secure the best conditions of work for their shops, while brigadiers represent the case of their brigades in just the same way as the forewoman of the *lampovaya* is expected to defend her collective.[20] The similarity in the position of male and female collectives is strikingly illustrated by the fact that even the locally famous brigade of development workers, one of the long-standing members of which was quoted above, was unable, as the quotation revealed, to resist being broken up when management decreed it. His brigade may well have been *splochennyi* and *skolochennyi*, but its very existence was revealed to be dependent on the whim of management.[21]

Thus, the system of representation by line managers is a general feature of the life of the mine, and male workers are equally dependent on the advocacy of their brigadiers or foremen. In the mining shops the brigadiers – who, aside from the duty (*zven'evoi*) brigadiers responsible for particular shifts rather than the brigade as a whole, are nearly always elected – have a very significant place in the structure of the shop and play a crucial role in negotiations with management over output and payment levels.[22] The structure varies slightly between production and development shops – in the former, one brigadier is responsible for all the face-workers, while in the development shops, there are a number of brigadiers responsible for different parts of the production process (for example, in a development shop there would be a brigadier elected by the explosives team) – but the role of the brigadiers is very similar. Most importantly, both attend the monthly planning meetings convened by the chief engineer to discuss output targets and payment for the month ahead. The role of

[20] In underground shops the foreman is responsible for health and safety, while the organisation of work is the responsibility of the brigadier. It is the brigadier rather than the foreman to whom workers turn to as their 'representative'.

[21] The *lampovaya* may actually be less divided than miners' brigades because it is a homogeneous collective with few significant variations in grade and skill. In contrast to this there are wide variations in skill and pay among underground workers, from the highly skilled combine operators (MGVM) to the lowly 'underground mine workers' (GRP). Such differences do cause some resentment. For examples see Ashwin, 1996: 33.

[22] In most of the auxiliary shops, by contrast, the position of brigadier is usually more formal, and unelected. In this situation, workers take their concerns over issues such as payment and work organisation to the foremen or forewomen.

brigadiers at these meetings – which are usually convened on a shop-by-shop basis – is to argue the case of their brigade: to agree an achievable plan with the administration; to ensure that they have the materials they require to meet it; and to agree a favourable level of payment for the brigade. This is an on-going war of attrition, as the following comment from the former head of the planning department illustrates. She acknowledged that the input of the brigadiers into the planning meetings was important because they knew the conditions 'on the ground', but at the same time she felt it was necessary to guard against their influence by restricting the information they received about concessions granted to other shops:

> If, for example, the chief engineer decides there is a particular problem in one shop he might say, 'OK, this month we'll give you a plan for eighty rather than a hundred metres or eighty rather than a hundred tonnes.' Well, if another shop hears that they'll want their plan reduced too. It's human nature. I think it's better to deal with people separately, so that people know as little about what is going on in other shops as possible. That way everything is much easier.

The brigadiers gave a similar picture of continual struggle. With regard to materials, for example, one of the brigadiers from a development shop claimed:

> The job of the brigadier is to demand the material that the brigade needs to fulfil the plan – the pipes, metal and so on, so that the brigade can do its job. We have to make sure that we actually get all the materials that we should according to the plan – this doesn't happen as a matter of course. We often don't get them.

The endemic tensions within the production process are thus channelled upwards through the brigadiers, who try to get the best deal that they can for workers. Workers reported that, in contrast to the shop trade union committees, their brigadiers looked after their interests. The following, for example, was a typical assessment of a worker from a development brigade:

> [The brigadier's] one of us. He'll come and help us out when we're working. He'll realise, for example, if we need another worker and he'll take one from somewhere else. He can raise the coefficient of those who work well a little bit and cut it from those who don't.

In production brigades the system is the same: as one *groz* (face-worker) put it, 'he [the brigadier] looks after our interests. He notes who works well and who works badly, and in accordance with this will raise or reduce [their] coefficient.'[23] In addition to this, the brigadiers will also stand up for individual workers who fall foul of the administration. As one of the brigadiers from a development shop explained, 'We look after pay-related issues. We sort it out with the shop chief if a good worker for some reason did not turn up for work;

[23] Interestingly, miners never complain of injustice in the administration of this system: the brigadier is 'one of us' and is not seen to use his discretion unfairly.

we get involved if someone is punished unjustly.' Here, the mentality of the trade union is discernible – brigadiers will defend *good* workers – but none the less, the brigadiers are in general far less equivocal than the union and certainly play a far more important role in handling conflicts within mining shops.

Can the system stand the strain?

Soviet enterprises have always been plagued by conflicts over supplies of materials required for production, work speeds and pay, but the economic disruption caused by economic reform has exacerbated all these tensions. At present, the conflict within the mine can be channelled up the mine hierarchy, but it is questionable how long it can be contained within this framework as the financial difficulties of the mine deepen. These mean that managers are both less able and less willing to grant concessions to workers, while the workers have more grievances than ever.

Cracks are beginning to appear within the system, as is apparent from conversations with brigadiers and miners. For example, the deals that the administration makes with the brigadiers with regard to payment cannot always be honoured as a result of the mine's financial problems. This obviously undermines the negotiation process between the brigadiers and the administration. As one face-worker commented of the planning meetings:

> A year ago we used to conclude a written agreement. It had to be signed by the brigadier, the shop chief and the [mine] president. And then the brigadier could demand our money. Now we're asking to have this system reinstated. Because now they can promise to pay us a lot, and then say that the mine hasn't got any money and pay us less. But if we had a written contract then we could demand our money.

The reinstitution of written contracts, however, though it would help resolve conflicts over back-pay, would not solve the problem that the mine is living from hand to mouth and that it is often simply unable to pay wages on time. And, likewise, there is no easy answer to the shortage of investment funds, which have an influence both on working conditions – because of the dangers and discomfort associated with the resultant 'make do and mend' approach to production – and on wages, because shortages of necessary materials mean that workers are unable to meet the plan. Although there were always shortages in the past, these have been exacerbated in the present period. As one miner noted: 'There is a problem with machinery, there is a constant lack of money. Either there's not a complex, or there's no conveyors. For example, in shop eight – they've put a complex in, but there are no conveyors. . . . It always works out like this here.' In this situation, it is becoming increasingly difficult for brigadiers to achieve a favourable deal for their brigades.

This does cause a good deal of tension but, in line with the argument of the last chapter, this is diffused somewhat by workers' adherence to a 'one enterprise' vision. Mineworkers generally consider that the mine's problems are caused by government policy towards the coal industry, and this only strengthens the idea that the solution is for the mine to pull together to weather the storm. This can be seen clearly in the following commentary from a miner on the way that the bargaining power of the brigadier was being eroded in the present period:

> He [the brigadier] does his best for us, because he's a worker just like us. If he knows that we won't fulfil the plan that they want to give us then he'll try to reduce it. But now because of non-payment [of subsidies and on the part of customers] he can't reduce the plan. And we try to fulfil the plan.

Workers try to fulfil the plan not only for the sake of their wage levels, but also for the sake of the enterprise: non-fulfilment of the plan is far more serious in an era when everyone is acutely aware that mine closures are being planned. Similarly, in auxiliary shops late payment of wages and shortages of materials are a constant cause of complaint, but generally this is directed at the government rather than the mine administration.

This situation can certainly continue for some time to come. But if managers begin to seek more authoritarian routes out of the crisis by cracking down on workplace discipline – and this process appears to be beginning at Taldym – the informal system of representation by line managers may begin to break down. Moreover, the present system has the potential to act as a force for fragmentation: as competition for scarce resources intensifies, there is a risk that the enterprise will be consumed by a destructive 'each shop for itself' struggle for survival. To avoid such a scenario, it is quite likely that enterprise managers will want to strengthen the authority of line managers, so that rather than channelling demands both from managers to workers and vice versa they would restrict themselves to the former. In such a situation workers would be left without even the 'representation' they currently enjoy.[24]

How would workers respond in such a situation? Since Daria Nikolaevna effectively redefined her role within the *lampovaya* from that of representative of the collective to representative of management, events in the *lampovaya* provide some preliminary answers to this question. First, nothing happened quickly, because of the problems of self-organisation within collectives discussed above. Workers did, however, begin to redefine the limits of the notional collective in their own minds. Since Daria Nikolaevna had, in the eyes of the collective, abrogated her responsibilities as forewoman, she became cast in the role of outsider, as a member of the hostile forces of the mine administration; she was no longer a member of the collective, although in this case it was not excluded that a different forewoman could be readmitted into its ranks. This could clearly be seen from the fact that the forewoman became the target of exactly the same

[24] Workers from a variety of collectives reported that it was again becoming more difficult for them to remove their managers.

sort of criticism usually reserved for those perceived as being part of the 'them' of the mine administration, rather than the 'us' of the collective. She came under increasingly strict scrutiny by workers accustomed to the feeling of being cheated by 'them'. Workers became very curious about the level of her salary, for example, and managed to find out through informal contacts in the accounts department how much she earned. They were horrified: she was said to earn just over twice their wage.[25] The forewoman was also accused of defrauding workers whenever she had the chance and of using the mine's resources for her own gain. In short, she was seen to be reproducing in microcosm all the sins of the mine administration. This can be seen clearly in the following comment:

> She is not honest. For example, the mine administration gave the *lampovaya* a delicate white Japanese kettle, the like of which we'd never seen before. She, without saying a word to anyone, just took it home with her. Of course, everyone knew but no one said anything. All of us would have liked to take that kettle home with us – what, am I not a human being too? Wouldn't I like a nice kettle too? I nearly said something – but why create unnecessary trouble for yourself?

The question posed here 'am I not a human being too?' clearly marks the forewoman out as being on the wrong side of the 'them and us' divide in this worker's mind: as revealed in the last chapter, some of the workers' most bitter denunciations of management concern the fact that the latter arrogate to themselves the right to live in a civilised manner and treat workers as a subhuman category. Similarly, the forewoman was felt, like other members of management, to be defrauding the mine for her own gain: 'She ordered some clear plastic to cover our *grafik* books. But she ordered twice as much as she needed for that. I suppose she used the rest on her allotment to make her polytunnels.' Such criticisms revealed that Daria Nikolaevna had become associated in the workers' minds with the mine administration, although this did not, as discussed above, lead them to change their expectations of the norms of behaviour expected from line management in general.

In the case of a wider strategy to introduce a more authoritarian model of line management, however, the possibility of re-election would be removed in order to liberate managers from populist pressures. Workers would therefore no longer have the option of seeking the perfect parent figure for their collective and would be forced to live with the manager imposed on them from above. Could this lead them to develop an increased consciousness of their interests as workers? Evidence from other enterprises does suggest that changes in the status and role of line management can result in increased conflict, and perhaps even a greater tendency towards collective action on the part of

[25] This level of differentiation was a new phenomenon: the forewoman had aroused curiosity when she began to hide her wage slip. The former *master*, Anna Petrovna, was as shocked by the extent of differentiation as anyone else.

workers. For example, Galina Monousova reports an incident at a Moscow watch-making enterprise, which she calls 'Device', where the sharp increase in pay differentials between workers and their line managers led to a situation in which, in the words of one foreman at the enterprise, 'we were forced to hide the list of wages, distrust arose among workers. We began to lose the manageability of the collective' (Monousova, 1996: 177). This discontent eventually manifested itself in a strike and the removal of the director of the enterprise by a vote of the labour collective. After the strike, however, which brought limited gains, workers, according to Monousova, began to feel nostalgic for the authoritarian style of the old director. The paternalist mentality is deeply ingrained, and even if, referring back to the example of Taldym, the current discipline drive does result in a more authoritarian model of line management,[26] there is no guarantee that workers would not initially hope that the situation could be improved under a new director, just as the workers at 'Device' did: the post-privatisation history of the mine would suggest that this is quite likely.

The way workers relate to 'the collective' is complex. At the very lowest level – the level of the brigade – there is a culture of co-operation and mutual assistance, but this is based on the organisation of work and does not readily extend into collective defence or representation. Representation is left up to line managers, who 'personify' the immediate work collective: the collective itself has no organisational force without its figurehead. This system was developed in the communist era when the global planning order, which defined the common interests of the labour collective, was replicated within the enterprise, and the 'unity' and interests of the immediate work collective as a productive unit took priority over the particular interests of workers. This system has survived the collapse of the central planning system; no alternative mechanisms of worker representation have developed. Some of the workers' needs can be met in this way – they can get 'a lick of paint' for their workplaces – but only when these do not directly contradict the interests of the administration. For example, brigadiers from production or development shops can attempt to ensure that their brigades have the materials they require to meet the plan and thus maintain wage levels, but they can do little about health and safety problems if their resolution will interfere with production. The collectivity of the immediate work group is thus not an expression of workers' self-organisation: the work group has a defined place within the mine hierarchy and representation by line managers is an established informal procedure which, for the moment, counters the collectivist potential of the division of the enterprise into

[26] This is a big 'if' in any case. The difficulties of Soviet and post-Soviet production mean that management is highly reliant on the informal co-operation of workers. 'Working to rule' in this environment has the potential to cause enormous disruption. Thus, there are strong pressures on managers not to behave as straight-down-the-line representatives of the administration. Fear of unemployment may, of course, begin to erode workers' capacity for informal, individualised resistance, but, as discussed in the previous chapter, this is as yet not the crucial determinant of workers' behaviour.

'collectives'. The system of representation by line managers is not without its tensions and fissures, but though it is presently subject to strain it has not yet broken down; conflict within the enterprise can still be channelled up the enterprise hierarchy. This leaves the trade union and its shop trade union committees untouched at the side-lines: increasing tension exerts no pressure on the union to reform, but rather places management structures under strain. Thus, workers remain dependent on the protection accorded to them by their managers. When they are let down they do begin to see their managers as members of the administration rather than as 'representatives', but, for the moment, workers resolve this dilemma by reversing the process of representation and appealing to the director to protect them from the unjust line manager.

8

Surviving the transition: individual and collective strategies

The previous chapters have provided an analysis of why the conflict unleashed by transition has neither resulted in a renewal of the old trade unions, nor prompted workers to form their own organisations outside the official structures, but they have not developed a systematic account of precisely how workers *are* responding to the privations of the reform era. This chapter considers workers' reaction to change from a different angle: instead of asking why workers are not organising, it analyses the nature of their survival strategies during the transition period and the political implications of these. For although workers do not constitute an organised presence on the political stage, their responses to reform, be they individual or collective, nevertheless play a crucial role in the post-communist recomposition of the Russian state and economy.

The argument developed here is that workers adopt complementary strategies in the face of reform. On the one hand, they are far from indifferent to the loss of security provided by the social guarantees of the past, or to the destruction of the collective institutions of social and welfare provision. But rather than seeking to build a new relationship between individual and collective, in which the workers would take control of their collective institutions, they remain locked into the alienated forms of symbolic collectivism inherited from the past, treating the collective as a resource imposed from above and seeking their salvation in a paternalistic leader who can promise to restore the security of the past, a salvation which the Kuzbass mineworkers seek through their support for the communists, and other brands of authoritarian leaders, at the local and national level.[1]

Meanwhile, as a complement to their search for collective salvation, workers are also very active in pursuing individual survival strategies. Such

[1] As mentioned earlier, in the years 1989–93 the mineworkers constituted a key element of the 'democratic' movement in Russia, but since 1993 support for the communists and nationalists has been growing in mining regions: by the 1995 election the once 'reformist' Kuzbass coal basin had become renowned as a communist stronghold, while the polar coalfield of Vorkuta, with its strong anti-communist traditions deriving from its gulag past, voted heavily for Zhirinovskii.

strategies, however, are pursued not through the social relations of the work collective, but through networks of family and friends, which are in most cases independent of the social relations of the immediate work collective. Enterprise collectivism is thus coming under pressure from below as workers increasingly look outside the enterprise for their survival – which in turn reduces the possibility of their mounting a collective response to transition and reinforces their dependence on authoritarian leaders. The complementary perspectives of individualism and 'alienated collectivism' adopted by workers are a key force in shaping the labour collective, and Russia, of the future.

Individual survival strategies

The story of the stunted development of collectivism in the *lampovaya* is just one example of the inability of mineworkers to realise the promise of 1989 by developing new kinds of self-organisation, a failure which can be seen on a wider scale in the failure to develop an effective independent trade unionism out of the workers' committees, and in the similar failure to realise the potential implicit in the form of privatisation in which ownership was transferred to the labour collective. It is not enough, however, to put this down to the system of line management and the culture of individualised informal relations within the enterprise. For 'the collective' (in all its forms) is also being redefined from below by the daily practice of workers as they increasingly look outside the mine to secure their survival. Although workers look to a paternalist for their salvation at work, they are increasingly engaging in various kinds of economic activity and building on networks of family and friends beyond the workplace.[2] Some of these opportunities existed in the Soviet period, but many have only opened up, or have only become legal, with the development of market relations, sometimes replacing facilities which were formerly provided by the enterprise.

The nature of the pressures faced by workers in the transition era was described in detail in Chapter 2: they are not paid on time; their savings have been wiped out by inflation; credit is drying up; housing is no longer free; and *sotskul'tbyt* is gradually being destroyed. Although, as argued in the previous chapter, workers' collectives are important to them, their immediate survival strategies in the face of such difficulties have been largely family-centred rather than based on networks at work. While in large urban enterprises workers are more likely to engage in paid secondary employment, often using skills acquired in their main jobs,[3] in mining settlements secondary employment largely takes the form of small-scale food production within the household. As described in Chapter 2, mineworkers have always had allotments or kitchen gardens

[2] Although such activities occur outside work, they can none the less be seen as the logical continuation of workers' attempts to cut individual deals within the enterprise: when their benefactors at the mine can no longer help them they are thrown back on their own resources.

[3] For more details see Donova and Varshavskaya, 1996.

attached to the self-built wooden houses of the 'private sector' and grown potatoes on the large tracts of land rented by the mine from local collective farms for its workers, and many of them kept cows, pigs and chickens in the past. Workers generally felt, however, that such activity had assumed greater importance during the transition period: as one fitter put it, 'what is 100 per cent true now is that we'd die without our allotments'. This sentiment was expressed in different forms in nearly every interview with workers in August 1996. Small-scale food production had also become a source of monetary income for many workers, who reported that, while in the past they had given their surplus away to friends, now they would usually sell it. Meanwhile, keeping cows is becoming increasingly popular, and the money to be made from selling milk more significant. One worker, for example, reported that she had kept cows all her working life, but in the past her wages had been worth far more than her milk money. Now she made a million roubles a month (at the time of the interview just under $200) from selling milk, while her monthly wage, when she received it, fluctuated between 600,000 and 800,000 roubles.

Although other forms of secondary employment are less common in Vishnovka,[4] the mine's financial difficulties definitely increase the relative importance of the household and family units in relation to work and the collective. Ironically, while the changes in shift system at Taldym can in one sense be seen as a 'collectivist' achievement, they also mean that both male and female workers spend fewer days in the month at work and thus have more time for their individual and family-based activities. Wage delays in particular cause workers gradually to disengage from the life of the mine: in the absence of wages, work obviously seems less important than production within the household or paid work outside the mine. In the most extreme cases, workers' outside interests come to dominate completely. This applies, for example, to those workers who buy themselves sick notes from doctors, and receive sick pay while involved in other work, most commonly trade. The trade union president reported that this was a serious problem at the mine: in August 1996, over 500 workers were on sick leave and he estimated that a substantial proportion of these had paid for fictitious sick notes. For this reason, the social security commission at the mine, which, after the collapse of the Party, had effectively ceased to operate, had been reactivated and delegations were again being sent to check up on 'sick' workers. Those who were found working on their dachas or trading at the bazaar were to be forced to return to work or to leave the mine. Other workers noticed that, even when they attended work regularly, those who had an regular income from their 'secondary' employ-

[4] Some workers with marketable skills, such as fitters and welders, are able to find paid work outside the mine, but this is not nearly as widespread as in urban areas. Since workers' food production enables them to survive they will usually only seek additional paid employment if faced with a specific financial need, a good example being the mine kindergarten teacher who had arranged a business buying meat in the Altai to sell in Osinniki because she needed to pay for her son's higher education.

ment were less committed to their work at the mine than they were in the past:

> There's a woman in my shop who goes to Novosibirsk and buys things there cheaply and then sells them here for more money – Vishnovka is a very expensive place. She goes there on her days off. There are lots of people like that. For example, Novikov's brother – he runs three or four kiosks and he still works at the mine. Pasha's old shop chief at the open cast mine had his own shop. People are beginning to have several jobs . . .[5]
>
> But the woman from my shop, she has less interest in her work. One of the other workers went to look for her at work the other day and she was off collecting berries during working hours. Well, in my book that's not a serious attitude to work. She's more interested in trading than work. There she gets money directly.

Although involvement in trade is far from universal, this example does illustrate that the mine is no longer as central to workers' survival as it once was. Meanwhile, young people are no longer being integrated within the labour collective: the 'mutual lack of love' between industrial enterprises and young people (Metalina, 1997) may be less prevalent in Vishnovka than in more urban parts of Russia, but it is none the less notable.

The post-retirement history of Anna Petrovna from the *lampovaya* in many ways exemplifies the nature of the contemporary changes in the Russian economy, and the social changes which accompany them. In November 1997, having decided she was sick of retirement, she found a job as a saucepan saleswoman for the Swiss firm Zepter. The cheapest set of saucepans available from the firm costs $999 in Russia; buyers make monthly payments and only receive the saucepans after they have made the last payment. Anna Petrovna is paid on commission and has a monthly target of $3,000 for the conclusion of sales contracts. Surprisingly enough, she has been able to meet her plan and became a team leader in March 1998. Her husband retired from the mine in December 1997 and now spends his time driving her round from client to client.[6] She gives demonstrations, extolling with great enthusiasm the virtues of the sophisticated saucepans which make it possible to cook without fat. This form of cooking, her company argues, will protect people from cancer, which is rife in the region. Anna Petrovna has become a true believer[7] and can talk of little other than the saucepans – even when she spoke of the hardship being suffered

[5] The extent of trading among mineworkers in Vishnovka is rather more limited than this would suggest: when workers talked about trading, they almost exclusively cited the same people as examples. This indicates that, rather than being commonplace, trading is, for the moment, still a visible exception to the norm within the labour collective.

[6] In a private moment she confided, 'women are stronger, they are the ones who sort things out here.'

[7] Her faith in the saucepans as a form of salvation from the unhealthiness of the local environment had an almost religious quality to it. Meanwhile, clients who are prepared to pay such prices for a set of saucepans also appear to see them as a talisman which will protect them from the surrounding chaos.

by her former colleagues in the *lampovaya*, she measured this on a Zepter-scale: 'They would love to buy the saucepans, but they will never be able to afford it.' Her (almost exclusively female) colleagues, she says, are mainly former teachers and health workers who have left their professions because of wage delays. In this way, the social structure of the region is being transformed: Anna Petrovna now earns more than the best-paid miners at Taldym.[8]

Along with the recognition that the mine no longer guarantees their security, workers also feel that work is now no longer the social focus and source of meaning in life it once was. This is not only a result of the greater importance attached to outside interests by workers; the attitude of management has also changed with the collapse of the Communist Party. They are no longer required to foster the integration of communist citizens through the promotion of 'the collective': they now have a more narrow concern with maintaining discipline and production levels. The change in atmosphere is well captured in the comments of this woman worker from the technical complex:

> Before they [management] kept us informed about what was going on. For example, they kept us up to date with all the *prikazy* [orders] at the mine, read them out to us at the *naryad* [the pre-shift task assignment meeting]. Now they don't. . . . According to the health and safety rules we have to have a *naryad.* We have to go along there and sign to say that we've had it. But before it was more interesting; we found out more there. And we could tell management what was wrong and what we needed. We talked there. . . . Before I used to get up and I was happy that I was going off to work. Now I wake up and I don't want to go to work: we don't even get paid. And the collective was better before. Before I'd even say that it was very good. But now people have turned nasty because of all the difficulties, and that influences relations.

These sentiments were echoed by the brigadier from the *zaryadnoe:*

> I liked it better in the communist period. Then we had the plan. We had to run round and meet it and if we did then we got rewarded – people had an interest in their work . . .
>
> The collective has changed. It was much more united before. Now it's fragmented [*razroznennyi*]. And people have become aggressive. . . . Before there was something to aspire to – there were the medal givings, the bonuses, it all meant something. I used to run round. I was in the shop Party committee, the shop trade union committee, the civil defence, I was bursting with energy and used to manage to get everything done. . . . Now if I'm honest I've lost interest in work a bit. For example, a wagon might come up dirty from the mine – whereas before I would have hosed it down, now I can't be bothered – let it stay dirty.

Institutions such as the plan, the Party and medals certainly had their downside: workers often complain of the divisiveness of the old system. But at

[8] It is interesting to note, however, that the company Zepter is in many ways aping the paternalist practices of Soviet past. For example, Anna Petrovna spoke with great excitement about the annual Zepter Ball held for the saleswomen in St Petersburg.

the same time they miss the sense of common endeavour and belonging that was fostered in the past.

The collective between the past and the future

Workers do not only regret the decline of the mine as a focus of communal sociability, they also have very concrete concerns about the decline of social provision and the threat to employment. For although the household has increased its relative importance as a productive unit, workers cannot survive without the mine: as one worker put it, 'We don't bake our own bread.' Indeed, the fact that many public services, such as hospitals and higher education institutes, have begun to charge fees means that workers are in many ways even more dependent on the mine as a source of monetary income than they were in the past. For example, the worker mentioned above who made a million a month through selling milk was still forced to apply to the mine for an urgent loan of five million roubles to pay for hospital care after her son was involved in a serious accident. Meanwhile, the mine still provides workers with a variety of other services such as free coal, loans of machinery for activities such as hay-making and holiday vouchers, as well as organising potato growing: as argued in Chapter 2, the domestic and enterprise economies are closely intertwined. Thus, even if they are beginning to expect less from the enterprise than they did in the past, workers' jobs are still very important to them and they continue to look to the mine for protection in times of acute need. Meanwhile, the continued importance of the productive sector of the economy more generally is nicely illustrated by the fact that the vast majority of Anna Petrovna's clients are managers from the mine: without the mine she, and a host of other petty traders, would have no business.[9]

After what has been said about their attachment to their collectives, it is no surprise that workers generally only seek work outside the mine out of necessity: they are not lured away from the dark, satanic pit by the intrinsic appeal of Russia's new 'service sector'.[10] As argued above, although Vishnovka does have some entrepreneurs among its mineworkers, regular involvement in paid work or trading is generally only sought when workers incur on-going financial commitments, such as bills for their children's education. Compared to the

[9] The irony of this was not lost on her. She was also well aware of the implications of the fact that managers were buying $999-saucepan sets while workers were not being paid. This seems to give strong support to workers' claims that managers pay themselves while leaving their subordinates without wages.

[10] Marina Ilyina's study of the *chelnoki* (shuttle traders) in Syktyvkar market (1996b) found that 70 per cent of respondents professed a desire to leave the shuttle trade, preferring either to open their own shop, work in their former profession, or indeed do any other work which would provide them with a good income. This was in spite of the fact that many of those she interviewed were successful, and were certainly making more money than they could elsewhere.

stimulation offered by the collective, trading in particular is seen as boring. As one woman worker confided:

> For me, it feels like a waste of time. I don't like it. You have to sit there wasting time waiting for people to come and buy things. I like to be doing something active. For example, this summer I sold some of my cucumbers. I sold them very cheaply, cheaper than everyone else, because I couldn't be bothered to sit there for ages – I wanted to get rid of them quickly. I suppose it meant that I sold my own work short.

Serious traders are required to travel regularly to Novosibirsk to buy cheap goods to sell. This is very tiring, especially when combined with work at the mine. One woman worker at the mine and her husband, a face worker, had discussed going into trade but had decided against it for this reason:

> We personally live on our pay. Volodya said – why don't I start trading? But I said no. I don't like the idea of it – I think it's a bit risky, although now we have the right to trade or do business if we want to. I could do it. I know how it's done. I've got a friend who was made redundant from a garment making factory and she now sells bras and knickers at Osinniki market. She sold so much before 8 March [Women's Day], she made a million. It's quite a stable business – it is the sort of thing people have to buy. I could do it too. I used to work as a shop assistant, so I have experience. But I'm scared that I would get too sucked into it and would be forever going off to Novosibirsk. I'd be travelling all the time and I'd forget about my children. The house would be a mess.[11]

Moreover, such 'unproductive' labour is still stigmatised and unfavourably compared with the honest productive labour of mineworkers. Indeed, some workers talked about trading as if it was some form of prostitution, as the following exchange between two women workers, discussing an acquaintance who had become a trader, graphically illustrates:

> WORKER 1: Good on her. She retired from the mine and now she trades at the market there. I see her there all the time. She seems to be doing well. I couldn't do it though. I just couldn't bring myself to . . . I don't think any of the girls from the *lampovaya* do it.
>
> WORKER 2: I think it's probably just the first time that's hard. Once you've done it once probably it's easy – you just get on with it after that.

Thus, while the mine is now less central to workers' lives than it once was, the alternatives to work in the mine are not viewed as attractive.[12] Individual

[11] Those who do make regular trips to Novosibirsk confirm that it is indeed very tiring, far more so than work at the mine. Through the change in their *grafik* women in particular have managed to achieve what they see as a desirable balance between home and work at the mine.

[12] After Anna Petrovna overcame her 'first-time' nerves she began to enjoy her new job. Unlike those who remain at the mine, her value system has noticeably altered: in June 1998 she confided that she now felt ill every time she returned to the *lampovaya* and she did not know how she had been able to work there for so long. Now, in contrast to the past, she saw 'clean work' as superior to the manual work at the mine.

strategies are an important means of surviving the transition, but in most cases they still only constitute a supplement to the declining support provided by the mine. Nevertheless, even as such they are gradually transforming the nature of the local economy.

Alienated collectivism

Since the mine is still important to workers, it is not accurate to portray the labour collective as simply being eroded by the individualising pressures of the market economy: individual survival strategies are complemented by the search for collective salvation within the context of the labour collective. But, as the preceding chapters have shown, the Soviet system of surplus appropriation and the internal dynamics of the enterprise to which this gave rise constituted a barrier to the collective expression of workers' specific interests. These barriers have not been removed by transition: the enterprise still constitutes a supplicatory unity, while *blat* is remains a major feature of the social administration of the enterprise. For this reason, workers' collective aspirations take the alienated form of dependence on the figure of the paternalist, who is supposed to insulate the labour collective from the chilly neo-liberal winds. The situation described in the *lampovaya* is thus replicated at a mine level: although the workers do not generally attempt to defend their position through organised activity, they do attempt to address their problems by hunting for that elusive strong leader, capable of guiding the mine through the troubled waters of transition.

This desire for protection is collectivist in the sense that it recognises that the fate of individuals is inextricably linked to the fate of the community of the labour collective. It should be stressed, however, that workers' aspirations only really find expression in the individual act of voting. Workers' lack of organisation means that even when they are gathered together at all-mine meetings to make decisions they do so as an aggregate of individuals rather than as a collective. The individualised character of supposedly collective deliberation is well captured by this miner's description of the conduct of meetings at the mine:[13]

> For example, a former director stood for president [of the mine]. . . . He had forbidden someone, let's say, financial help. He didn't help another person build a house. . . . Or he arrived, for instance, at the *lampovaya* and a miner came [up from the pit] early. He says to the *lampovshchitsy*: note down that he's early. . . . Then the miner doesn't get his bonus. That's that. Everyone has their revenge. . . . Because of every private vengeance – he did badly by that one, that one and that one – they won't vote for him. But for the enterprise as a whole –

[13] This was part of a long lamentation about the political life of the mine, prompted by a question about the trade union. In his discussion of meetings the speaker did not distinguish between trade union meetings and meetings of the AO. In this he is typical: workers have a hazy conception of the different functions of the various types of mine meeting.

> as an industrialist he's good, he's clever, he's got connections, and as an industrialist he understands everything. He can keep the mine together, keep it on the right track. But people don't understand . . .
>
> This is how it works here: if I've got the gift of the gab, I've got a little bit of understanding, I can speak in front of a meeting, don't get embarrassed, can get across my point of view to the men – they'll get behind me. After me there'll be the next speaker – he'll give his point of view. And then they won't go with me but against me. The third one will speak, he'll again speak for me and the men will go along with him. See, they sway here and there: they don't have their own opinion. Even when we're discussing some kind of proposal. They are that stupid that on the same question that they can put their hands up to vote both for and against. Before my very eyes someone has voted in favour of something, and then someone else has spoken against it and he already puts his hand up to vote against.

Although this account is highly coloured, it is not inaccurate. Workers 'don't have their own opinion' because they are relating to the questions as individuals who have, as was established in previous chapters, a shifting sense of collective identification. Thus, workers themselves do not organise to promote one or other candidate in mine elections, they respond to the campaigns run by the management factions.[14] Despite their lack of organisation, however, their aspirations on behalf of the collective are in fact very similar and constitute an aggregate pressure for protection.

This could be seen in the contest for post of director at the mine at the end of 1995, in which workers were united in their desire for a more adequate protector. The previous director was a democratic populist whose main virtue was considered to be the fact that he did not put up barriers between himself and the workers. The following compliments – the first from a miner, and the second from a *lampovshchitsa* – are typical:

> I like him. For example, he knows I live near him and if he sees me he'll offer me a lift home. He'll give a few of us a lift at the same time. . . . He's attentive to people; he tries to help.
>
> He doesn't raise himself above others. You can go to him with any question. If he's got the time he will definitely see you. You can ask him anything and he'll give you an answer.

The former director had been elected because he was a popular shop chief who was seen as being 'close to the people': this qualification, it was felt, would ensure that he would look after the workers. He did, as the quotations above illustrate, often do personal favours for individual workers, and he did not

[14] Crowley (1997) claims that 'miners appropriated the formal democratic rituals of the old system and infused them with new meaning and substance' (203). But this is what miners at Taldym have signally failed to do: in-mine democracy remains a largely formal process in which managers call the shots. See Clarke and Kabalina (1995) for an account of the mechanisms through which managers retain control at enterprises in other branches, in which workers have been given democratic rights through privatisation to the labour collective.

attempt to tighten discipline at the mine. But the labour collective as a whole suffered under his populist regime: production, and hence wages, fell relative to past levels, and the mine fell into debt. Moreover, because of his (popular) lowly status as a former shop chief this director did not have the connections required to extract resources from the concern and from Moscow: wage arrears began growing as soon as he was elected. He attempted to compensate for this by making concessions to pacify particular groups of workers, the result of which was a growing belief that he had allowed discipline to collapse at the same time as failing to protect the majority of workers.

The new candidate for director, a deeply reassuring paternalist and former communist-era mayor of a nearby town, directly appealed to workers' desire for security.[15] The most important plank of his programme was the fact that he had good connections which he would exploit to the full in order to gain resources for the mine. As he put it in an interview:

> At the moment I can help the mine through my personal contacts. They are crucial. If you don't have them, you could spend two days outside an office door waiting to be seen. Whereas I can just pick up the phone. For example, we've got a new complex. We haven't paid for it yet. We are going to pay for it in coal over several months. I was able to arrange that because they know and trust me. They know that I'll keep my word.
>
> While these possibilities exist I will use them. I will get all I can from my contacts . . . I am trying to resist the policies of Moscow. They told me to shut down Vishnovskii *raion* [part of the mine], I invested money in it; they told me to close one of the faces, I put money into it.

He also promised, as this quotation indicates, to revive the 'Vishnovskii *raion*', and to find the money to finish building the blocks of flats, the construction of which had been frozen under the former director. The points workers most often recalled from his programme were his (well-advertised) good contacts with the local coal concern, and his promise to finish the flats (something, as Chapter 2 indicated, he duly did). Having had a taste of life under the leadership of an 'outsider' workers were keen to return to an experienced pair of communist hands.

What does this contest say about the political implications of workers' 'collectivist' strategy? First, yet again, it highlights the structural constraints on collective organisation imposed by the post-Soviet production system. Because the labour collective does still constitute a supplicatory unity in the face of the 'outside world', it is very difficult for workers to break out of the form of politics in which their interests are represented by the managers who do battle on their behalf, whether within the mine, or, in the case of the director, outside it. Again, the example of 1989 is relevant here. In this case, workers did try to develop their own demands, but their struggle for a decent standard of living none the

[15] This alternation between populist and authoritarian directors has been a common feature of the post-*perestroika* era, particularly in the coal-mining industry.

less ended up being fused with the struggle of the managers to forward the claims of the industry and region in Moscow.

Second, the alienated form of workers' collective action severely limits its political impact: such action ends up reproducing precisely those structures which caused the problem which promoted the action in the first place. Acquiring a more capable director can sometimes improve the position of the enterprise, but it does not challenge workers' dependence on the beneficence of the paternalist director. And, as the discussion of the labour collective revealed, they are very often disappointed in the performance of their putative protectors.

As has been stressed at the end of each chapter in this section, however, the Russian economy is changing. Although so far the structures within which conflict has been contained have been largely preserved, the system is subject to increasing strain. 'Alienated collectivism' has its limits and contradictions. First, it is becoming increasingly hard for managers to protect their enterprises, no matter how good their contacts. The new director of Taldym may be far better connected than his predecessor, but that has not insulated the mine from the problem of wage arrears, which has only intensified during his time in office. And he is aware that contacts do not offer a long-term solution to the mine's problems:

> If government policy doesn't change then the mine will slowly die. That is definite. But I am going to do all I can – use every contact – to make sure it's the very last to close. . . . But there may come a time in the future when all the leaks in the system are shut off. I might phone up my friend and he might say, 'I trust you, you're a good friend, but there's nothing I can do to help you.' Life is life and economics is economics.

Second, economics being economics, the community of interests between workers and managers is not without its contradictions. For the director is unlikely to deliver the sort of protection that workers are looking for: their aspirations are actually very different from those of management. Both want to preserve particular elements of the 'Soviet way of life', but they are not nostalgic for the same aspects of the authoritarian paternalism of the traditional Soviet enterprise. While mine management focuses on authoritarianism and looks back to the days of order and discipline when the boss's word was law, the workers focus on the paternalistic features of the traditional enterprise and look back to the security that was guaranteed by the collective. Workers' desire for protection does express itself in the search for a 'strong leader', but they are none the less deeply ambivalent about such strength when its effect is to reduce their autonomy and increase the intensity of their work. The return to communism at Taldym is by no means, therefore, an unambiguous yearning for the restoration of the past, but rather expresses two very different perspectives on the future development of a new Russia. Neither management nor workers want to renounce the opportunities and independence which they believe that they have achieved with the destruction of the Soviet system and the development of a market economy. But for management, order and discipline will provide the framework within which the enterprise can adjust to the demands of the changing economy, while for the workers, paternalism will provide the means by which they can be partially protected from the ravages of the market.

Workers' desire to be defended by strong leaders during the transition period does not only manifest itself at the mine level. It is also apparent in their attitude to national politics, in particular in the Kuzbass, where support for the Communist Party is strong. While they aspire to the security of collectivism, however, they can only envisage the realisation of this through the benevolence of a line manager, a paternalist director or the president of Russia, who personify the collective at each level. This alienated collectivism is exerting a major influence on the development of Russian politics. On the one hand, workers scurry around, too absorbed in their daily struggle for survival to organise, while on the other their combined aspirations further the development of authoritarian-paternalist strands of politics in the enterprise, the region and the country as a whole. Thus, ironically, workers' conscious attempts to improve their position do little to transform the structures which define their subordination: their dependence on increasingly authoritarian leaders only makes them more vulnerable. Meanwhile, however, their improvised, individual survival strategies are actually transforming economic relations and, regardless of their intentions, helping to foster market relations in Russia.

9

Conclusion

> We need a *khozyain* [at the mine] . . . Russia needs a *khozyain*. (Worker, *kotel'naya*, August 1996)

> If we get a good leader, the whole of the Kuzbass will explode. (Former production engineer, June 1998)

The 1996 presidential elections did nothing to satisfy workers' craving for a protective paternalist, for a 'firm, but fair' *khozyain*. Instead, they were left to cope with the ravages of a 'bad father' (indeed a quintessential bad father with a drink problem). In August 1996 the trade union president was unusually laconic: Yeltsin's election victory meant that the mine's difficulties would only deepen. His sole remaining hope was that, 'there *must* be a social explosion now. . . . People just can't endure this forever.' But there has been no explosion. The Kuzbass miners' blockade of the Trans-Siberian railway in May 1998 was a desperate attempt to draw attention to the plight of the region, but, though it brought temporary release, it had no impact on the overall direction of reform. No one interviewed at the mine in June 1998 thought the protest had resolved the problems; no one believed that the government would keep to its plan of paying off back wages; everyone was deeply despondent and pessimistic about the future. The dominant feeling was one of powerlessness; even the once-ebullient trade union president could not see a way forward, and wanted to retire from the fray to concentrate on protecting his own family.

This book has shown the many mechanisms through which conflict is contained within the enterprise. The result is that workers do not have their own organisations and have a shifting sense of where their interests lie. In this situation, as the previous chapter argued, their collective action tends to reproduce existing relations of domination rather than transforming them, because in place of self-organisation workers look to a leader to resolve their problems. As noted in Chapter 4, the trade union president, while wishing for a social explosion, had also warned darkly that when Russians ran out of patience 'then it's a nightmare'. In a sense he is right to be nervous of workers 'breaking out'. In the absence of workers' self-organisation this is not likely to bring any form of liberation: as the 1989 miners' strike shows, in the midst of truly spontaneous and largely unorganised outbursts new hierarchies can emerge with surprising rapidity.

But collective organisation is not easy, even among miners. Dennis *et al.* argued that:

> Solidarity, despite the division into interest groups among the miners in a given pit, is a very strongly developed characteristic of social relations in mining; it is a characteristic engendered by the nature and organisation of coal mining: it is a characteristic that has been given added strength as a result of the high degree of integration in mining villages. A miners' first loyalty is to his 'mates'. ([1956] 1969: 79)

Similarly, Bulmer (1975) stressed the solidarity of mining communities in his ideal type. Russian miners certainly express a rhetorical solidarity with each other, and this finds limited realisation at the level of the work group. Meanwhile, Russian miners have been a great deal more active than their counterparts in most other industries. But they are not organised.[1] What the foregoing chapters have shown is that solidarity and organisation do not emerge naturally: some circumstances may be more conducive than others, and the Russian environment is unquestionably not conducive, but even in more favourable climates organisations have to be created through purposeful political action. As Beynon and Austrin argued with regard to the establishment of a miners' union in the Durham coal field:

> In arduous and often brutal working and living conditions, the miners combined to form a stable union, supporting it with regular subscriptions, and invested in banners which symbolised their involvement. The village as a 'community' became a vital aspect of their identities, but this development was not simply inevitable and should not be taken for granted: solidarity had to be *built*. (1994: 364)

Why has the establishment of effective workers' organisations proved so difficult in post-communist Russia? The failure of the NPG to organise miners outside the official union was partly, as mentioned above, a matter of strategy. But the fact that Rosugleprof has had no more success with its 'reform from above' highlights the structural constraints on the development of organisations representing workers' interests in post-communist Russia. Most importantly, the position of the enterprise within the post-Soviet economy means that trade unions end up representing the interests of the labour collective as a 'supplicatory unity' rather than the specific interests of workers. This pattern of relations is difficult to escape both because the labour collective does have a

[1] Crowley (1997) seeks to explain what he calls 'miners' dramatic radicalisation' (199) and argues that 'miners are organised, ready for the coming battle that unemployment is sure to bring' (205). Sadly, this optimistic analysis is not supported by the facts. The mine closure programme is now under way, and the response of miners so far has been weak, unco-ordinated and ineffective.

common, and presently overriding, interest in survival, and because trade union committees, given their distance from workers, are dependent on managers to tolerate their existence. Courting the workers is therefore a risk which, if it does not pay off, can lead to the expulsion of the union from the enterprise. As discussed in Chapter 4, even the Taldym trade union president's limited independence has angered the mine director, who is now working with the NPG at the mine. He has allowed the NPG president to become a full-time official, he has given him a car (while confiscating that of the Rosugleprof committee), and he signed the 1998 collective agreement with the NPG rather than the Rosugleprof committee at the mine. Despite these advantages, the membership of the NPG at the mine has only increased moderately to 294 out of 2,600 mine employees. It is clear that with this level of membership the union will be in no better position to resist the pressure for collaboration with management than was its former communist rival.[2] The president of the NPG at the mine, like the mine president of Rosugleprof, does not lack good intentions,[3] but his freedom of action is severely constrained.

It is not simply a question of structure constraining the action of well-meaning agents, however: the Rosugleprof trade union president at Taldym does not even consider working in a different way. So far the essential features which have prevented the trade unions from representing the specific interests of workers remain precariously in place. Even as conditions change, however, there is no guarantee that the union, or the workers, will adapt with any rapidity to their new environment: as Bourdieu has pointed out, the tendency of groups to persist in their ways results partly from the fact that they are composed of 'individuals with durable dispositions which can outlive the economic and social conditions in which they were produced' (Bourdieu, 1990: 62). At present there is no indication that the union is about to transform the practice which leads workers to condemn it as 'a pocket trade union'. Change will only come about through protracted and complex interactions between union officers, workers and management in the context of a 'reforming' economic and political system.

[2] Vladimir Ilyin has shown how, after its initial heroic period of frequent strike action, the NPG in Vorkuta was faced with the mundane question of how to prevent its members defecting back to the official unions. It chose to do this through the distribution of barter goods, which in turn 'demanded that they normalise their relations with the administration since such distribution is in essence an honorary assignment to carry out administrative functions' (Ilyin, 1996b: 72).

[3] Talking about his duties in May 1998 the NPG president commented, 'I defend workers. Workers are not defended, they need a good trade union. A trade union should not give out vouchers and products – a trade union should defend your rights – pay, health and safety, and ensure you are paid on time. . . . I think we should say to the director, "It's your problem. You're the director, you should make sure workers are paid. How, I can't tell you because it's not my job, but you should do it. And if you can't someone else should take your place". . . . If I advise workers to be patient what kind of defence am I?' This contrasts with the position of the Rosugleprof president, who does effectively tell workers to be patient, although it is notable that the NPG president still sees a more effective director as the answer to workers' problems.

The other possibility which has been considered in foregoing chapters is a process of change initiated from below. First, it has been argued that workers are not exerting pressure on the union to reform because conflict within the enterprise is channelled up the enterprise hierarchy of line management, rather than being dealt with by the shop trade union committees, which primarily function as part of the enterprise welfare infrastructure. Moreover, the relation of the mine trade union committee to management also structures the way in which the shop trade union committees relate to workers: the committees are not encouraged to represent workers' interests, and worker involvement at shop level is not promoted by the union. Thus, increasing tension within the enterprise has not provided a catalyst for 'reform from below', but has instead placed management structures under strain.

Second, workers are not organising outside the union. They are 'getting by' as best they can, increasingly relying on production within the household and networks outside work, while at the same time clinging to the protection of the labour collective. They recognise that this entity is deeply divided but, at a time when its very existence is under threat, conflicts over distribution within the enterprise are eclipsed by the struggle to secure resources from outside agencies. In the communist era, the unity of interests of the labour collective in relation to the external environment did not make it easy for workers to define and defend their specific interests. This was an objective problem related to the structure of the Soviet economic system which found its expression in workers' shifting subjective identification. The barriers to the subjective realisation of a common class position were thus not linguistic but structural: the argument of Rossman (1997) and Davies (1997) regarding the ability of workers to use communist categories to criticise the powers that be has been supported by the evidence of this study. In the transition era, the 'production pact' (Clarke *et al.*, 1993: 99) between workers and managers, based on their common interest in maximising resources available to the enterprise and in attaining plan targets, has not been broken. Indeed, in some ways it has strengthened: now the enterprise constitutes a supplicatory unity defined by its common interest in survival.

Despite the 'production pact', under communism the endemic disputes over distribution within the enterprise did foster an oppositional mentality in which workers saw the world as divided between an underprivileged 'us' and a corrupt and parasitic 'them'. But the apparent unity of the 'simple workers' concealed the fact that the discretionary nature of the distribution and disciplinary regime within the enterprise obliged workers to become involved in the very deal making and system of *blat* which so enraged them. The centrality of individualised informal relations in the administration of the enterprise not only meant that workers became personally dependent on their superiors, it also ensured that their reflex response to any problem was to seek an individual rather than a collective solution. This has structured workers' response to transition: for example, rather than mounting protests against wage delays when they were first employed as a management strategy, workers tightened their belts and attempted to find individual solutions to the problems of living without wages.[4] The government and employers were thus free to allow wage arrears to mount.

There is some evidence that money is replacing the informal 'currency' of *blat*, but the culture of individualised relations extended beyond the distribution of the benefits in kind which are now being axed by enterprises. It was pervasive and deeply rooted: it will not be eroded overnight.[5] Moreover, those at the mine who have attempted to expose corruption, such as the former chair of the mine audit commission, have not received workers' support. Despite the fact that such people are speaking out against precisely the practices which enrage workers most, those who attempt actively to oppose the culture of mutual favours and protection which persists at the enterprise are seen as naive: they are quickly reminded that this is how Russia works and that it is senseless to try to swim against the tide.[6] There is also no evidence that the end of central administrative allocation is reducing opportunities for corruption: in the absence of the guiding hand of the Communist Party all limits on the use of position for personal gain have been removed. Meanwhile, now that Big Brother

[4] Galina Monousova (1996) provides another good example of this in her case study of the Moscow enterprise she calls 'Pizza'. Here, the management decided to restrict the amount of annual leave its predominantly female workforce could take during the summer, despite the fact that according to the law women with school-age children have the right to take their vacations whenever they like. Although the main production workers were opposed to the change, however, they did not mount a collective response to it: as one of them put it, 'we are afraid to speak openly. Some are waiting for an apartment, some for an allotment' (170). In interviews most of the women claimed that they would resolve the problem individually by going on the sick list in the summer.

[5] Kathryn Hendley has made a similar point with regard to legal reform in post-Soviet Russia. She argues, 'With the introduction of market mechanisms . . . economic actors have adapted their old tactics of using patron–client networks rather than shifting to a reliance on law. Their behavioural patterns or routines are supported by a set of informal norms that are well understood by all but remain unwritten. . . . Changing these well-established behavioural patterns will require more than passing new laws and/or reforming or creating new institutions' (Hendley, 1997: 246).

[6] In 1995, the chair of the mine audit commission presented evidence of serious violations on the part of management to a meeting of the AO, but a clear majority of worker-shareholders refused to support his no confidence motion. Anna Petrovna explained why: 'You see, these things always happen. For example, . . . the *raion* mine technical inspector – the director has to do him any favours that he asks. Any. Otherwise he can close down a shop, or even the whole mine. So the director gives him a car, builds him a house, builds him a garage. Does whatever is necessary. The inspectors are very low paid – this is how they make their money. The director didn't say this openly at the meeting – but he didn't need to – the miners understand. But for some reason Grachev [the chair of the audit commission] can't understand it. Perhaps it is because the director refused to give his brother a loan. . . . The workers don't really blame the director for problems like this, because they realise that it is just part of the system.' The fact that Anna Petrovna assumed that Grachev had a private motive for his concern over corruption illustrates the fact that such opposition to the system is seen as an aberration requiring explanation. Assuming that his action was a form of 'private vengeance' was a way of normalising Grachev's behaviour.

has been banished, the protection of powerful patrons is becoming more valuable than ever.[7]

Alongside this culture of individual dependence a system of collective dependency continues to be reproduced. Work collectives rely on their line managers to represent their interests within the enterprise in the same way as workers expect the director to do battle on behalf of the labour collective in the 'outside world'. Thus, neither the collectivity of the labour collective nor the close relations of work groups have formed a basis for workers' self-organisation. Instead, workers' search for collective salvation expresses itself in an alienated collectivism in which their aspirations are focused on the figure of the leader. Their political response to the crisis has thus been to turn to strong leaders who claim to be capable of affording them the protection for which they yearn. When they are let down by these 'representatives' their solution is not to challenge their dependence, but to seek a replacement for the inadequate protector.

How far do these conclusions specifically refer to mining, or indeed to Taldym? This depends on whether the structural forms which shaped the processes which have been identified exist in other branches and enterprises. The arguments developed above rely on a general characterisation of the nature of the production relations which existed within the communist system and their development in the post-Soviet era, and in this sense it is clear that I do not consider the processes identified to have a purely local character. This assertion does not merely depend on the argument that the Soviet system was highly centralised and 'monolithic' in character. It is based on empirical evidence that similar processes do indeed occur within other enterprises in other branches of industry, in particular evidence drawn from the collaborative research programme of the Centre for Comparative Labour Studies at Warwick University and the Institute for Comparative Labour Relations Research in Russia, which has involved longitudinal case studies of a variety of different types of industrial enterprise in Moscow, Kemerovo *oblast'*, Samara, Syktyvkar and Vorkuta. Individual chapters have frequently referred to the published results of this research, and the arguments of the book have also been informed by the discussions conducted at research seminars of this collaborative programme and by individual discussions with members of the group.

The material gathered from other former state industrial enterprises (although not new commercial enterprises, which were established in different conditions) shows that the structural similarities between post-Soviet enterprises in terms of management hierarchies, paternalist social provision, the

[7] In wider society the weakness of the state has implied a quasi-feudal fracturing of authority in which people are forced to rely on private power to enforce contracts, to protect themselves and to establish businesses: as Shlapentokh has argued 'probably nowhere has the notion of a "roof" [protection] had such an immense impact on political and economic processes as in Russia, where it has become an institution successfully competing with the official state' (Shlapentokh, 1996, cited in Hendley 1997: 243).

legacy of the state planning system and the position of the former communist trade unions have given rise to similar forms of social relations within labour collectives. For example, my account of the nature of informal relations within the enterprise is similar to that discovered by Sergei Alasheev in industrial enterprises in Samara (Alasheev, 1995a), Marina Kiblitskaya in a Moscow Metro train repair plant (Kiblitskaya, 1995), and Marina Ilyina and Vladimir Ilyin in a Syktyvkar passenger transport enterprise (Ilyin and Ilyina, 1996). Taldym does have specific features, which I have drawn attention to at various points, but these do not invalidate the key elements of the arguments outlined above.

In terms of Taldym's specificity, one of its particular characteristics constitutes the reason why it was chosen as the site for the case study: the fact that it has an active trade union committee which is affiliated to the union which is considered to be one of the most 'reformed' and militant of the former communist trade unions. This only serves to highlight the degree to which the constraints on union reform are structural in nature. It is the case, as was argued in Chapter 1, that the coal industry's dependence on subsidies leads the trade unions in this branch to adopt a particular role in lobbying for resources. This gives miners' unions a focus for their campaigns: the state. The same is true of the health and education unions. And it is not coincidental that these unions are the most active in the post-communist era: lacking a focus, many of the other branch unions have had problems defining their role during transition. Thus, unions in branches no longer dependent on the state have not developed new strategies and modes of action distinct from those of unions in state-dependent sectors: rather, these unions have generally languished. It may be that where multinationals take over, or a 'real owner' decides to drop all the last vestiges of paternalism, individual enterprise unions may become radicalised. Research is needed into the forms of industrial relations which are developing in such contexts, but for the moment the cases where unions have transformed their mode of operation remain very much the exception.[8] Moreover, even in privatised, non-subsidised enterprises there is a tendency for 'one enterprise' politics to persist, since such enterprises are generally constituted as besieged entities rather than surplus-value-extracting, profit-making concerns. In such cases the director remains a father or mother figure, expected to protect the remnants of the labour collective from the chill winds of neo-liberalism.

Another distinguishing feature of Taldym is its status as a closed AO. Workers' search for the perfect paternalist occurs within the context of the mine electoral system laid down in the constitution of this AO: 'democracy' at the mine has reinforced workers' expectation of directoral defence from the

[8] A representative of the International Federation of Chemical, Energy, Mine and General Workers' Unions (ICEM) claimed that he knew of one case in the chemical industry where an attempt by a multinational to exclude an enterprise union had resulted in the union's radicalisation. But without further examination it is difficult to judge such claims: it is possible to refer to Taldym's union as 'radical', but the union has not transformed its accustomed mode of operation.

outside world. The right to elect the director is not confined to this type of privatised enterprise, however, and aspirations are not only focused on the director when it is possible to replace him or her. The most important way in which the situation at Taldym differs from that at other enterprises is that the form of privatisation opted for has institutionalised a particular balance of power between workers and managers. Workers at Taldym are fortunate in working at a relatively 'good' mine, at which their influence over the divided administration has spared them some of the privations generally associated with reform. The workers have preserved shift systems which allow them to combine home and work in a way that suits them (which is increasingly important at a time when production within the household is vital for survival), and so far management has done its best to avoid making any compulsory redundancies. Thus, the labour collective's series of 'protectors', though considered by its members to have been deeply inadequate, has actually delivered them more benefits than more secure management teams at other enterprises.

Taldym is in a better situation than the average Soviet enterprise, but as a labour collective it is still in a state of slow decay. It can no longer provide for workers and they are increasingly thrown back on their own resources in order to secure their physical survival. This disintegration of one of the key institutions of Soviet society has profound consequences. For although the labour collective, as the means through which workers were integrated into the communist system, was in many ways a repressive institution, it was also the forum in which workers defined their identity. The ideas of the dignity of labour and the value of collectivism which were fostered within it were, as has been demonstrated, double-edged. On the one hand, the emphasis placed on the importance of pride in work was used as a means to stimulate workers to meet the plan, often against all the odds, but at the same time such ideas formed the basis on which workers defined themselves against their 'unproductive' managers. Meanwhile, the collective was a locus of control, but it was also a community, a focus of sociability, which contained within it an albeit unrealised germ of liberation. There was some prospect that the emancipatory potential contained within the labour collective could have been realised after the collapse of communism, but as the account of the *lampovaya*'s history in the post-communist period demonstrated, this did not occur. As the financial situation of the mine deteriorates, there is less and less prospect of such a development: as argued above, now the only priority is survival. The unrealised potential of reform is well captured by the comments of the senior teacher at one of the mine's kindergartens, which, at the time of the interview in August 1996, was about to be transferred to the local authority, under the auspices of which it faced a highly uncertain future:

> There have been lots of negative aspects of reform, but it also has positive aspects. Before there was just one programme which all kindergartens were obliged to follow. Now the new tendency is to develop the personality of every child. The word *lichnost'* [personality] has arrived in our language. We now try

> to bring up well-rounded ['many-sided'] children. I'd even say we try to develop their souls. For seventy years this didn't happen. Now we think about the emotional development of the children and of the collective.
>
> The children are much more comfortable here now because of the changes in the system. Before it was awful – if the inspector came round and one of the children couldn't sing the Soviet national anthem, then you got it. It meant that the teachers weren't paying sufficient attention to the development of patriotic spirit in the young. We were terrified of the inspectors. They used to come round looking for faults. Now we're not scared of them, they come round to help us. Things have unconditionally improved in that regard. Now we are their colleagues, whereas before we were subordinates.
>
> The work is much more interesting now – we can choose any programme we like, the only problem being, as I said, that there's no money for retraining. We need it. Basically, we can take everything they taught us before, put a big cross beside it and forget it. Before politics came first. When I remember how scared we were of that history of the CPSU . . .
>
> But in our collective the women were always kind and motherly to the children, and they haven't turned nasty because of their financial problems. We can say that the children don't suffer. We have also got some more equipment in the last few years, but generally we live much worse.

Unfortunately, the arrival of the word *lichnost'* in the language coincided with the launch of a structural adjustment programme which imposed financial constraints so stringent as to inhibit most forms of individual and collective self-realisation. Indeed, the fate of kindergartens is a good example of this. These were adjuncts of the traditional Soviet labour collective which had their faults, but at the same time their destruction represents a huge loss for the communities which they served, especially at a time when it has finally become possible to redefine them as nurturing rather than controlling institutions.

Is it too late for Russian workers to influence the course of transition through collective self-organisation? This book has shown that workers socialised in the Soviet era have a considerable attachment to collectivist principles, even though this has not yet been realised in collective self-organisation. The question of whether the obstacles which have prevented this can be overcome is a complex one: as has been shown, structural barriers to reform exist at all levels of the trade union hierarchy, while workers are also subjectively and objectively constrained from organising collectively. It is therefore difficult to identify the most likely source of change. At the same time, however, the pressures on workers are intensifying, and, as has been shown, in extreme circumstances the workers do attempt to defend themselves. So far such reactions have been short-lived and largely unorganised, but the frequency of such protests is increasing: the Novokuznetsk–Tashtagol railway was blocked by miners for twenty-five days in July 1998, while at least five other miners' blockades occurred in Kemerovo *oblast'* during the same month. In order to transform spontaneous action into something more significant, however, activists will have to learn from the mistakes of the strikes of 1989 and 1991. That is, they will have to concentrate on building democratic organisations, rather than relying on patronage from the political authorities. This is not easy, but it is now more necessary than ever. For the main political problem in contemporary Russia at all levels is the lack of democratic accountability. This creates almost unlimited scope for corruption and puts no pressure on politicians, officials and managers to consider the interests of ordinary people: it is no wonder that the leaders in whom workers place their trust always disappoint them. The establishment of an effective trade union or workers' movement would be a huge step towards supplanting the post-Soviet cult of leadership with a culture of accountability in Russian politics.[9] It is therefore to be hoped that, rather than seeking short-term political gain, the new generation of workers' leaders will direct their attention to the more arduous task of building democratic organisations.[10]

But the environment is becoming steadily less conducive to workers' organisation. For, while the social form of the labour collective inhibited workers' organisation, its decomposition presages the destruction of the communality of experience which formed the basis of Russian workers' identity. In Vishnovka, this pushes workers towards an impoverished peasant existence,[11] while in larger towns the gradual demise of enterprises plunges workers into a 'world-in-between' in which 'they live on handouts from friends and relatives, from subsistence agriculture, from occasional casual labour, petty trading or petty crime' (Clarke, 1996a: 75). Down river from Vishnovka in the small village of Kuzedeeva the closure of the settlement's main employer, a small toy factory, has meant that the majority of those who were once its workers are now living off the land. Such developments not only mean a huge drop in living standards for those affected, they also destroy the basis of the erstwhile workers' self-identity. Since decaying labour collectives take on few young workers, and young people are not attracted to enterprises that pay wages several months late, the Russian working class, which in the past was defined by its common (if contradictory) existence within the labour collective, is fragmenting. And this, as the experience of Western democracies in recent years testifies, poses a whole series of new problems in terms of forging solidarities and forming workers' organisations.

[9] It is not only Russians workers who perpetuate this obsession with leaders: Western governments and the international financial institutions have also been guilty of identifying the future of Russian democracy with the future of President Yeltsin and whoever is deemed to be the most convincing young neo-liberal of the moment. The pitfalls of this approach are now becoming painfully apparent.

[10] A new generation of miners' leaders is currently beginning to emerge. On 8 September 1998, *Nezavisimaya gazeta* tipped the Vorkuta miner Viktor Semenov, who heads the All-Russian strike committee, as the logical nucleus for the next social revolution in Russia (*RFE/RL Newsline*, 2, 174, Part I, 9 September 1998). Typically, the paper envisaged Semenov in a Leninist role, but there exists at least the possibility that he and other such leaders will eschew this path in favour of creating organisations accountable to the rank and file.

[11] As has been argued, the economies of the household and enterprise were closely intertwined, so the closure of the mine really would force unemployed mineworkers into extreme poverty.

References

Abercrombie, N. and Hill, S. (1976), 'Paternalism and Patronage', *British Journal of Sociology*, 27, 413–29.

Alasheev, S. (1995a), 'Informal Relations in the Soviet System of Production', in S. Clarke (ed.), *Management and Industry in Russia: Formal and Informal Relations in the Period of Transition*, Aldershot, Edward Elgar, 28–68.

Alasheev, S. (1995b), 'On a Particular Kind of Love and the Specificity of Soviet Production', in S. Clarke (ed.), *Management and Industry in Russia: Formal and Informal Relations in the Period of Transition*, Aldershot, Edward Elgar, 69–98.

Arnot, B. (1988), *Controlling Soviet Labour: Experimental Change from Brezhnev to Gorbachev*, Basingstoke and London, Macmillan.

Ashwin, S. (1994), 'Trade Unions After the Fall of Communism in Eastern Europe and Russia', *Journal of Area Studies*, 5, 137–52.

Ashwin, S. (1995), 'Russia's Official Trade Unions: Renewal or Redundancy?', *Industrial Relations Journal*, 26:3, 192–203.

Ashwin, S. (1996), 'Forms of Collectivity in a Non-Monetary Society', *Sociology*, 30:1, 21–39.

Ashwin, S. and Bowers, E. (1997), 'Do Russian Women Want to Work?', in M. Buckley (ed.), *Post-Soviet Women: From the Baltic to Central Asia*, Cambridge, Cambridge University Press, 21–37.

Batstone, E. (1988), *The Reform of Workplace Industrial Relations. Theory, Myth and Evidence*, Oxford, Clarendon Press.

Beynon, H. and Austrin, T. (1994), *Masters and Servants: Class and Patronage in the Making of a Labour Organisation*, London, Rivers Oram Press.

Bialer, S. (1980), *Stalin's Successors: Leadership, Stability and Change in the Soviet Union*, New York, Cambridge University Press.

Borisov, V. and Clarke, S. (1996), 'The Russian Miners' Strike of February 1996', *Capital and Class*, 59, 23–30.

Borisov, V., Bizyukova, V. and Burnyshev, K. (1996), 'Conflict in a Coal-Mining Enterprise: A Case Study of Sudzhenskaya Mine', in S. Clarke (ed.), *Labour Relations in Transition: Wages, Employment and Industrial Conflict in Russia*, Aldershot, Edward Elgar, 234–74.

Bourdieu, P. (1990), *The Logic of Practice*, Cambridge, Polity Press.

Breslauer, G. (1984), 'On the Adaptability of Soviet Welfare-State Authoritarianism', in E. Hoffman and R. Laird (eds), *The Soviet Polity in the Modern Era*, New York, Aldine.

Bulmer, M. (1975), 'Sociological Models of the Mining Community', *Sociological Review*, 23, 61–92.

Burawoy, M. and Lukács, J. (1992), *The Radiant Past: Ideology and Reality in Hungary's Road to Capitalism*, Chicago, University of Chicago Press.

Burnyshev, K. and Pulyaeva, O. (1996), 'Ugol'naya kompaniya v usloviyakh sokrashcheniya gospoddezhki (Sotsial'no-upravlencheskie aspekt' razvitiya)', mimeo.

Carr, E. H. (1952), *The Bolshevik Revolution, 1917–1923. Volume 2*, London, Macmillan.

Chalmers, N. (1989), *Industrial Relations in Japan: The Peripheral Workforce*, London, Routledge.

Clarke, S. (1996a), 'Structural Adjustment without Mass Unemployment? Lessons from Russia', in Institute for Comparative Labour Relations Research and Centre for Comparative Labour Studies, *The Restructuring of Employment and the Formation of a Labour Market in Russia*, Warwick, Centre for Comparative Labour Studies, 35–82.

Clarke, S. (1996b), *The Development of Industrial Relations in Russia. Report for the ILO Task Force on Industrial Relations*, Warwick, Centre for Comparative Labour Studies.

Clarke, S. (ed.) (1996c), *The Russian Industrial Enterprise in Transition: Case Studies*, Aldershot, Edward Elgar.

Clarke, S. (1998), 'S. Crowley, Hot Coal, Cold Steel: Russian and Ukrainian Workers from the End of the Soviet Union to the Post-Communist Transformations' (review), *Europe–Asia Studies*, 50:4, 721–3.

Clarke, S. and Kabalina, V. (1995), 'Privatisation and the Struggle for Control of the Enterprise', in D. Lane (ed.), *Russia in Transition: Politics, Privatisation and Inequality*, New York and Harlow, Longman, 142–58.

Clarke, S., Ashwin, S. and Borisov, V. (1997), 'The Non-Payment of Wages in Russia', report prepared for the ICFTU.

Clarke, S., Fairbrother, P. and Borisov, V. (1995), *The Workers' Movement in Russia*, Aldershot, Edward Elgar.

Clarke, S., Fairbrother, P., Borisov, V. and Bizyukov, P. (1994), 'The Privatisation of Industrial Enterprises in Russia: Four Case-Studies', *Europe–Asia Studies*, 46:2, 179–214.

Clarke, S., Fairbrother, P., Burawoy, M. and Krotov, P. (1993), *What About the Workers? Workers and the Transition to Capitalism in Russia*, London, Verso.

Cook, L. J. (1993), *The Soviet Social Contract and Why it Failed: Welfare Policy and Workers' Politics From Brezhnev to Yeltsin*, Cambridge, Mass., and London, Harvard University Press.

Connor, W. D. (1988), *Socialism's Dilemmas: State and Society in the Soviet Bloc*, New York, Columbia University Press.

Connor, W. D. (1991), *The Accidental Proletariat: Workers, Politics and Crisis in Gorbachev's Russia*, Princeton, N.J., Princeton University Press.

Conquest, R. (1967), *Industrial Workers in the USSR*, London, Bodley Head.

Crouch, C. (ed.) (1979), *State and Economy in Contemporary* Capitalism, London, Croom Helm.

Crowley, S. (1997), *Hot Coal, Cold Steel: Russian and Ukrainian Workers from the End of the Soviet Union to the Post-Communist Transformations*, Ann Arbor, University of Michigan Press.

Davies, S. (1997), '"Us against Them": Social Identity in Soviet Russia, 1934–41', *Russian Review*, 56, 70–89.

Davis, J. and James, S. (1986), 'Women from the Valleys Turn Activist', in V. Seddon (ed.), *The Cutting Edge. Women and the Pit Strike*, London, Lawrence and Wishart, 16–29.

Dennis, N., Henriques, F. and Slaughter, C. [1956] (1969), *Coal Is Our Life*, London, Tavistock.

Dolzhenok, V. and Sal'nikov, I. (1982), *Shakhta 'Alardinskaya'*, Kemerovo, Kemerovskoe knizhnoe izdatel'stvo.

Donova, I. (1996), 'Wage Systems in Pioneers of Privatisation', in S. Clarke (ed.), *Labour Relations in Transition: Wages, Employment and Industrial Conflict in Russia*, Aldershot, Edward Elgar, 41–62.

Donova, I. and Varshavskaya, L. (1996), 'Secondary Employment of Employees of Industrial Enterprises', in Institute for Comparative Labour Relations Research and Centre for Comparative Labour Studies, *The Restructuring of Employment and the Formation of a Labour Market in Russia*, Warwick, Centre for Comparative Labour Studies, 135–48.

Dore, R. P. (1973), *British Factory – Japanese Factory*, Berkeley, Calif. and Los Angeles, University of California Press.

Fairbrother, P. and Waddington, J. (1990), 'The Politics of Trade Unionism: Evidence, Policy and Theory', *Capital and Class*, 41, 15–56.

Filtzer, D. (1992), *Soviet Workers and De-Stalinization: The Formation of the Modern System of Soviet Production Relations, 1953–1964*, Cambridge, Cambridge University Press.

Filtzer, D. (1994), *Soviet Workers and the Collapse of Perestroika: The Soviet Labour Process and Gorbachev's Reforms, 1985–1991*, Cambridge, Cambridge University Press.

Friedgut, T. and Siegelbaum, L. (1990), 'Perestroika from Below: The Soviet Miners' Strike and its Aftermath', *New Left Review*, 181, 5–32.

Fulcher, J. (1987), 'Labour Movement Theory Versus Corporatism: Social Democracy in Sweden', *Sociology*, 21:2, 231–52.

Godson, J. (1981), 'The Role of the Trade Unions', in L. Schapiro, and J. Godson, *The Soviet Worker. Illusions and Realities*, London and Basingstoke, Macmillan, 106–29.

Gordon, L. and Klopov, E. (1991), 'The Independent Workers' Movement in the USSR', in T. Tchertvernina, L. Gordon and E. Klopov, *Labour Market Trends and Labour Market Policy for Transition. Wage Developments and Industrial Relations in the USSR*, Paris, ILO, OECD, 33–64.

Hauslohner, P. (1987), 'Gorbachev's Social Contract', *Soviet Economy*, 3:1, 54–89.

Haynes, V. and Semyonova, O. (eds) (1979), *Workers Against the Gulag*, London, Pluto Press.

Hendley, K. (1997), 'Legal Development in Post-Soviet Russia', *Post-Soviet Affairs*, 13:3, 228–51.

Hewett, E. (1988), *Reforming the Soviet Economy: Equality Versus Efficiency*, Washington, D.C., Brookings Institution.

Hoffman, D. (1994), *Peasant Metropolis: Social Identities in Moscow, 1929–1941*, Ithaca, N.Y. and London, Cornell University Press.

Hough, J. F. (1979), 'Policy Making and the Worker', in A. Kahan and B. Ruble (eds), *Industrial Labour in the USSR*, New York and Oxford, Pergamon Press, 367–96.

Hyman, R. (1975), *Industrial Relations: A Marxist Introduction*, London and Basingstoke, Macmillan.

Hyman, R. (1989), 'The Sickness of British Trade Unionism: Is There a Cure?', in R. Hyman, *The Political Economy of Industrial Relations: Theory and Practice in a Cold Climate*, Basingstoke, Macmillan, 166–87.

ICFTU (1992a), *The Trade Union Asset Question in Russia*, Brussels, ICFTU.

ICFTU (1992b), *Report on the CIS, October* 1992, Brussels, ICFTU.

ICFTU (1998), *Report on the Russian Miners' Strike*, Brussels, ICFTU.

Ilyin, V. (1996a), 'Social Contradictions and Conflicts in State Enterprises in the Transition Period' in S. Clarke, (ed.), *Conflict and Change in the Russian Industrial Enterprise*, Aldershot, Edward Elgar, 34–64.

Ilyin, V. (1996b), 'Russian Trade Unions and the Management Apparatus in the Transition Period', in S. Clarke (ed.), *Conflict and Change in the Russian Industrial Enterprise*, Aldershot, Edward Elgar, 65–106.

Ilyin, V. and Ilyina, M. (1996), 'On the Buses: Management Dynamics in a Passenger Transport Enterprise', in S. Clarke (ed.), *The Russian Enterprise in Transition: Case Studies*, Aldershot, Edward Elgar, 333–92.

Ilyina, M. (1996a), 'Foremen: An Ethnographic Investigation', in S. Clarke (ed.), *Labour Relations in Transition: Wages, Employment and Industrial Conflict in Russia*, Aldershot, Edward Elgar, 63–81.

Ilyina, M. (1996b), 'Shrikhi k portretu "chelnoka": Keis-stadi veshchevogo rynka Syktyvkara', paper presented to ISITO seminar, Syktyvkar, September 1996.

Jacobi, O., Keller, B. and Müller-Jentsch, W. (1992), 'Germany: Codetermining the Future?', in A. Ferner and R. Hyman (eds), *Industrial Relations in the New Europe*, Oxford, Blackwell, 218–69.

John, A. V. (1980), *By the Sweat of their Brow: Women Workers at Victorian Coal Mines*, London, Croom Helm.

Kabalina, V., Monousova, G. and Vedeneeva, V. (1996), 'Lenkon: Capitalism and Scientology', in S. Clarke (ed.), *The Russian Enterprise in Transition: Case Studies*, Aldershot, Edward Elgar, 125–247.

Kelly, J. (1988), *Trade Unions and Socialist Politics*, London, Verso.

Kharkhordin, O. (1996), '"A System of Responsible Dependency": The Collective as an Object of Knowledge and Action' (draft of Ph.D. chapter), UC Berkeley.

Kiblitskaya, M. (1995), 'We Didn't Make the Plan', in S. Clarke (ed.), *Management and Industry in Russia: Formal and Informal Relations in the Period of Transition*, Aldershot, Edward Elgar, 198–223.

Kirk, N. (1994), 'History, Language, Ideas and Post-Modernism: A Materialist View', *Social History*, 19:2, 221–40.

Kjellberg, A. (1992), 'Sweden: Can the Swedish Model Survive?', in A. Ferner and R. Hyman (eds), *Industrial Relations in the New Europe*, Oxford, Blackwell, 88–142.

Kotkin, S. (1994), 'Coercion and Identity: Workers' Lives in Stalin's Showcase City', in L. Siegelbaum and R. Suny (eds), *Making Workers Soviet: Power, Class and Identity*, Ithaca, N.Y. and London, Cornell University Press, 274–310.

Kotkin, S. (1995), *Magnetic Mountain: Stalinism as a Civilisation*, Berkeley, Calif., Los Angeles and London, University of California Press.

Kozina, I. (1996), 'Changes in the Social Organisation of an Industrial Enterprise', in S. Clarke (ed.), *Labour Relations in Transition: Wages, Employment and Industrial Conflict in Russia*, Aldershot, Edward Elgar, 146–59.

Kozina, I. (1997), 'The Mechanism of Staff Reduction and the Consequences for the Social Organisation of the Collective during Transition', paper presented to ISITO seminar, Moscow, May 1997.

Kozina, I. and Borisov, V. (1996), 'The Changing Status of Workers in the Enterprise', in

S. Clarke (ed.), *Conflict and Change in the Russian Industrial Enterprise*, Aldershot, Edward Elgar, 136–61.

Kumazawa, M. (1996), *Portraits of the Japanese Workplace: Labour Movements, Workers, and Managers*, Boulder, Colo., Westview Press.

Kumazawa, M. and Yamada, J. (1989), 'Jobs and Skills Under the Lifelong *Nenko* Employment Practice', in S. Wood (ed.), *The Transformation of Work*, London, Unwin Hyman, 103–26.

Kuusinen, O. (ed.) (1961), *Fundamentals of Marxism–Leninism*, London, Lawrence and Wishart.

Lane, D. (1985), *Soviet Economy and Society*, Oxford, Blackwell.

Lapidus, G. (1983), 'Social Trends', in R. Brynes (ed.), *After Brezhnev: Sources of Soviet Conduct in the 1980s*, Bloomington, Indiana University Press.

Lehmbruch, G. (1982), 'Introduction: Neo-Corporatism in Comparative Perspective', in G. Lehmbruch and P. C. Schmitter, *Patterns of Corporatist Policy Making*, London, Sage, 1–28.

Lenin, V. I. [1921] (1947), 'Preliminary Draft Resolution of the Tenth Congress of the Russian Communist Party on the Syndicalist and Anarchist Deviation', *The Essentials of Lenin. Volume 2*, London, Lawrence and Wishart, 683–6.

Lenin, V. I. [1922] (1947), 'The Role and Functions of Trade Unions Under the New Economic Policy', *The Essentials of Lenin. Volume 2*, London, Lawrence and Wishart, 760–9.

Madison, B. Q. (1979), 'Trade Unions and Social Welfare', in A. Kahan and B. Ruble (eds), *Industrial Labour in the USSR*, New York and Oxford, Pergamon Press, 85–115.

Mandel, D. (1991), *Perestroika and the Soviet People: Rebirth of the Labour Movement*, Montreal, Black Rose Press.

Mandel, D. (1993), *Rabotyagi: Perestroika and After Viewed from Below*, New York, Monthly Review.

Marsden, D. and Richardson, R. (1994), 'Performing for Pay? The Effects of Merit Pay on Motivation in a Public Service', *British Journal of Industrial Relations*, 32:2, 243–61.

Martin, R. and Fryer, R. H. (1973), *Redundancy and Paternalist Capitalism*, London, Allen and Unwin.

Marx, K. [1867] (1990), *Capital. Volume 1*, Harmondsworth, Penguin.

McAulay, M. (1969), *Labour Disputes in Soviet Russia 1957–65*, Oxford, Clarendon Press.

Metalina, T. (1996), 'Employment Policy in an Industrial Enterprise', in S. Clarke (ed.), *Labour Relations in Transition: Wages, Employment and Industrial Conflict in Russia*, Aldershot, Edward Elgar, 119–45.

Metalina, T. (1997), 'Young People and Industrial Enterprises', paper presented to ISITO semimar, Moscow, May 1997.

Michels, R. (1915), *Political Parties: A Sociological Study of the Oligarchical Tendencies of Modern Democracy*, Glencoe, Ill., Free Press.

Monousova, G. (1996), 'Gender Differentiation and Industrial Relations', in S. Clarke (ed.), *Conflict and Change in the Russian Industrial Enterprise*, Aldershot, Edward Elgar, 162–90.

Newby, H. (1977), 'Paternalism and Capitalism', in R. Scase (ed.), *Industrial Society: Class, Cleavage and Control*, London, Allen and Unwin.

Nove, A. (1989), 'Is there a Soviet Ruling Class?', in A. Giddens and D. Held (eds), *Classes,*

Power and Conflict: Classical and Contemporary Debates, Berkeley, Calif., University of California Press, 588–604.

Panitch, L. (1980), 'Recent Theorisations of Corporatism: Reflections on a Growth Industry', *British Journal of Sociology*, 31, 159–87.

Piatakov, A. (1962), 'Labour Administration by the State and Trade Unions in the USSR', *International Labour Review*, 85, 558–73.

Romanov, P. (1996), 'Moi zavod – moi dom', paper presented to conference on 'Gender Relations in Transition', Soros Foundation East–East Programme, Samara, Russia, 24–7 May.

Rose, R. (1994), *Getting by Without a Government: Everyday Life in a Stressful Society*, Glasgow, University of Strathclyde, Centre for the Study of Public Policy.

Rossman, J. (1997), 'The Teikovo Cotton Workers' Strike of April 1932: Class, Gender and Identity Politics in Stalin's Russia', *Russian Review*, 56, 44–69.

Ruble, B. (1981), *Soviet Trade Unions: Their Development in the 1970s*, Cambridge, Cambridge University Press.

Rutland, P. (1990), 'Labour Unrest and Movements in 1989 and 1990', *Soviet Economy*, 6, 354–84.

Samara Research Group (1995), 'Paternalism in the Russian Enterprise: Our Understanding', in S. Clarke (ed.), *Management and Industry in Russia: Formal and Informal Relations in the Period of Transition*, Aldershot, Edward Elgar, 128–39.

Schapiro, L. and Godson, J. (1981), *The Soviet Worker: From Lenin to Andropov*, Basingstoke, Macmillan.

Schmitter, P. C. (1984), 'Still the Century of Corporatism?', *Review of Politics*, 36, 85–131.

Seddon, V. (ed.) (1986), *The Cutting Edge. Women and the Pit Strike*, London, Lawrence and Wishart.

Shapiro, J. (1992), 'The Industrial Labour Force', in M. Buckley (ed.), *Perestroika and Soviet Women*, Cambridge, Cambridge University Press, 14–38.

Shlapentokh, V. (1996), 'Early Feudalism – The Best Parallel for Contemporary Russia?', *Europe–Asia Studies*, 48:3, 393–411.

Siegelbaum, L. and Suny, R. (1994), 'Class Backwards? In Search of the Soviet Working Class', in L. Siegelbaum and R. Suny (eds), *Making Workers Soviet: Power, Class and Identity*, Ithaca, N.Y. and London, Cornell University Press, 1–26.

Slider, D. (1987), 'The Brigade System in Soviet Industry: An Effort to Restructure the Labour Force', *Soviet Studies*, 39, 3, 135–55.

Smith, S. (1997), 'Russian Workers and the Politics of Social Identity', *Russian Review*, 56, 1–7.

Stedman Jones, G. (1983), *Languages of Class: Case Studies in English Working Class History, 1832–1982*, Cambridge, Cambridge University Press.

Taylor, S. (1986), 'Grub Up For The Miners!', in V. Seddon (ed.), *The Cutting Edge. Women and the Pit Strike*, London, Lawrence and Wishart.

Teague, E. (1989), 'Embryos of People's Power', *Report on the USSR*, 32, 1–4.

Ticktin, H. (1992), *Origins of the Crisis in the USSR: Essays on the Political Economy of a Disintegrating System*, Armonk, N.Y., M. E. Sharpe.

Varshavskaya, L. (1998), 'Sotsial'nye fenomen Sibirskoi "fazendi"', paper presented to ISITO seminar, Dubrovnik, September 1998.

Vedeneeva, V. (1995), 'Payment Systems and the Restructuring of Production Relations in Russia', in S. Clarke (ed.), *Management and Industry in Russia: Formal and Informal Relations in the Period of Transition*, Aldershot, Edward Elgar, 224–39.

Walder, A. (1986), *Communist Neo-Traditionalism*, Berkeley, Calif., University of California Press.

Warde, A. (1989), 'Industrial Discipline: Factory Regime and Politics in Lancaster', *Work, Employment and Society*, 3:1, 49–63.

Working Centre for Russian Economic Reform (1977), *Russian Economic Trends*, 1, London, Working Centre for Russian Economic Reform.

World Bank (1994), *Russian Federation. Restructuring the Coal Industry: Putting People First*, Washington, D.C., World Bank.

World Bank (1996), *From Plan to Market. Executive Summary. World Development Report 1996*, Washington, D.C., World Bank.

Zaslavsky, V. (1982), *The Neo-Stalinist State. Class, Ethnicity and Consensus in Soviet Society*, Brighton, Harvester Press.

Index

Note: 'n.' after a page reference indicates the number of a note on that page.